WORKING PAPERS NUMBER 4

Robert P. Scharlemann
Editor

ISSUES IN EVOLUTION, HISTORY AND PROGRESS

Lanham • New York • London

COMMITTEE ON COMPARATIVE STUDY
OF INDIVIDUAL AND SOCIETY

Copyright © 1990 by
University Press of America®, Inc.
4720 Boston Way
Lanham, Maryland 20706

3 Henrietta Street
London WC2E 8LU England

All rights reserved
Printed in the United States of America
British Cataloging in Publication Information Available

Co-published by arrangement with the
Committee on the Comparative Study of the
Individual and Society of the Center for Advanced Studies
at the University of Virginia

"Consciousness out of Context: Evolution, History,
Progress and the Post-Post-Industrial Society" and
"A Reply: Bearing the Bad News,"
copyrighted 1990 by Robin Fox

Library of Congress Cataloging-in-Publication Data

Issues in evolution, history, and progress.
p.   cm. — (Working papers / Committee on the
Comparative Study of the Individual and Society of the
Center for Advanced Studies at the University of Virginia ; no. 4)
1. Social evolution. 2. Progress. I. Series: Working papers
(University of Virginia. Committee on the Comparative Study of
the Individual and Society) ; no. 4.
GN360.I87   1990   303.44—dc20   89-38620 CIP AC

ISBN 0-8191-7638-9 (alk. paper)

  The paper used in this publication meets the minimum requirements of
American National Standard for Information Sciences—Permanence
of Paper for Printed Library Materials, ANSI Z39.48-1984.

CONTENTS

General Introduction	iv
Editor's Foreword	v
Consciousness out of Context: Evolution, History, Progress and the Post-Post-Industrial Society **Robin Fox**	1
Consciousness, Relativism and Utopia **Richard Handler**	39
Science and Human Nature **George Klosko**	49
A Commentary on Fox's Diagnosis of the Human Condition **Julian N. Hartt**	59
A Reply: Bearing the Bad News **Robin Fox**	78
Members of the Committee on the Comparative Study of the Individual and Society, 1988-89	87

GENERAL INTRODUCTION

The Committee on the Comparative Study of the Individual and Society occasionally publishes a volume of *Working Papers* on a theme or issue of sufficient interest to its members. Sometimes the Committee-sponsored programs - occasional lectures, panel discussions, and colloquia - may also provide a basis for issuing a working paper. The purpose of these publications is to encourage wide-ranging discussion of an issue within the University and outside. The readers of these papers are encouraged to share their ideas with members of the Committee contributing to a particular volume of the *Working Papers*.

This issue is again dedicated to Mr. W.D. Whitehead, Director of the Center for Advanced Studies and Dean of the Graduate School of the University of Virginia, who has supported with insight and catholicity the interdisciplinary work of the Committee during the last decade and has encouraged bringing out the *Working Papers* as a partial record of the Committee's deliberations.

EDITOR'S FOREWORD

Robin Fox's essay, "Consciousness out of Context," was presented for a panel discussion held at the University of Virginia in the Spring of 1988. As the replies of the three panelists indicate, it was a provocative essay both for them and also for the audience. We hope it will be so for the readers as well. Fox's thesis - if I may paraphrase it loosely - that the human species is biologically constituted in such a way as to make progress beyond a certain stage of organization unrealizable, and that what we regard as the progress of civilization is instead only a kind of oscillation between the two extremes of an impossible situation, is one that overturns many commonplaces of modern thought. Professor Fox has presented the thesis with clarity and vigor, and his respondents have been equally clear (and vigorous) in their comments upon it, so that there is no need for an editorial preview of the contents. The editor will, however, take this opportunity to thank all of the participants for their contributions and wish readers as much profit from these essays as was enjoyed by the participants and audience.

CONSCIOUSNESS OUT OF CONTEXT: EVOLUTION, HISTORY, PROGRESS AND THE POST-POST-INDUSTRIAL SOCIETY

Robin Fox
University Professor of Social Theory
Rutgers University

What is your opinion of Progress? Does it, for example
Exist? Is there ever progression without retrogression?
Therefore is it not true that mankind
Can more justly be said increasingly to Gress?

      Christopher Fry, *A Phoenix Too Frequent*

History, Stephen said, is a nightmare from which I am trying to awake.

      James Joyce, *Ulysses*

A trout is only as smart as he has to be: a fisherman is twice as smart as he needs to be.

      Old Fly-fishing Proverb

    Yes, Virginia, this is yet another "commentary on the human condition." But despite the ponderous overtones, this really is a serious tradition and one I am glad to belong to. But the conditions under which we now have to write such commentaries have drastically changed. Two things have altered our sense of ourselves radically: the daily prospect of total annihilation for our species, and our very recent awareness of our ancientness as a species, with all the implications this has for an understanding of ourselves.
    It is a sad irony that the latter crucial knowledge, which promises to transform our ideas of what we are and what we can hope for in

the future, has been acquired at exactly the time when we are threatening to make that future impossible. We are like someone who has been handed a great fortune along with instructions to commit suicide.

Reason, imagination, and violence today coexist in a way we can only try to analyze or express. But the two cultures of reason and imagination - the wrong basis for the antagonism between the humanities and the sciences - do not exist out there in the world; they only exist in our categorical reconstruction of it. Before Plato they were not sundered, but before Plato there was no science either. Yet Plato wanted the poets and artists out of the Republic because, as Eric Havelock (*Preface to Plato*) rightly observes, they laid claim to a rival truth - not just to a superior capacity for entertainment. So Plato argues for their banishment:

Οὐκοῦν δικαίως ἂν αὐτοῦ ἤδη ἐπιλαμβανοίμεθα, καὶ τιθεῖμεν ἀντίστροφον αὐτὸν τῷ ζωγράφῳ· καὶ γὰρ τῷ φαῦλα ποιεῖν πρὸς ἀλήθειαν ἔοικεν αὐτῷ, καὶ τῷ πρὸς ἕτερον τοιοῦτον ὁμιλεῖν τῆς ψυχῆς ἀλλὰ μὴ πρὸς τὸ βέλτιστον, καὶ ταύτῃ ὡμοίωται. καὶ οὕτως ἤδη ἂν ἐν δίκῃ οὐ παραδεχοίμεθα εἰς μέλλουσαν εὐνομεῖσθαι πόλιν, ὅτι τοῦτο ἐγείρει τῆς ψυχῆς καὶ τρέφει καὶ ἰσχυρὸν ποιῶν ἀπόλλυσι τὸ λογιστικόν, ὥσπερ ἐν πόλει ὅταν τις μοχθηροὺς ἐγκρατεῖς ποιῶν παραδιδῷ τὴν πόλιν, τοὺς δὲ χαριεστέρους φθείρῃ· ταὐτὸν καὶ τὸν μιμητικὸν ποιητὴν φήσομεν κακὴν πολιτείαν ἰδίᾳ ἑκάστου τῇ ψυχῇ ἐμποιεῖν, τῷ ἀνοήτῳ αὐτῆς χαριζόμενον καὶ οὔτε τὰ μείζω οὔτε τὰ ἐλάττω διαγιγνώσκοντι, ἀλλὰ τὰ αὐτὰ τοτὲ μὲν μεγάλα ἡγουμένῳ, τοτὲ δὲ σμικρά, εἴδωλα εἰδωλοποιοῦντα, τοῦ δὲ ἀληθοῦς πόρρω πάνυ ἀφεστῶτα.

Republic, book 10

This perhaps necessary divorce, however, like so much of the Platonic heritage, may yet cost us dearly. It may yet cost us everything. As Voltaire, in 1770, passionately declaimed:

O Platon tant admiré, j'ai peur que vois ne vous ayez conté que des fables, et que vous n'ayez jamais parlé qu'en sophismes. O Platon! vous avez fait bien plus de mal que vous ne croyez. Comment cela? me demandera-t-on: je ne le dirai pas.
Dictionnaire philosophique

Why wouldn't he say what the damage was? Perhaps because at that time it had not been fully assessed. Perhaps because Voltaire sensed the dangerous fragmentation that Plato had wrought, but had not seen its full effects. If that is the case, Voltaire was truly prophetic. And again there is a real irony, because when Plato wanted to express the great truth of truths, in the *Timaeus*, he resorted to the language of myth and poetry, and obviously meant us to take him seriously. (Which leads me to wonder if the mass of the Socratic dialogues are not perhaps some kind of send up - a vast satire on the misuse of reason.)

But the reintegration that our condition cries out for cannot be on a pre-Platonic basis: a tribal, heroic, oral basis. Our capacity for literate communication and self-annihilation has outrun that possibility. Yet we do not seem able to make the necessary imaginative/rational leap into the state of consciousness needed to handle this monstrous situation that we have ourselves created out of the debris of the Platonic sundering of reason and imagination.

Where must we start in trying to re-understand ourselves? It will be a basic contention here that we must condemn as deficient any commentary on the human condition that fails to take into account the ancientness of the species and the more than five-million years of natural selection that have molded the questionable end product that includes the commentators and their commentaries. That is to say, anything but an evolutionary view of modern man is going to be insufficient if its purpose is to calculate the possibilities of human survival. This rules out most of the governing paradigms of informed commentary, scientific, literary, political, and religious. In particular it rules out most of social theory as it now stands. To see why, let us look at one of the best examples of modern social theory; for what the author of this theory finds hard to handle, given its assumptions, turns out to be the cornerstone of an evolutionary understanding.

For Whom the Bell Tolls

I would recommend anyone to read Daniel Bell's brilliant analysis of trends in the contemporary world, *The Coming of Post-Industrial Society*. I marvel at the skill with which he analyzes the complexities of industrial society and its transformations. In particular he makes a good case for taking the "disjunction between culture and social structure" as the major agent of change. Roughly speaking, this means that ideas, values and knowledge get out of kilter with established social institutions, and when this gets disjunctive enough, a readjustment has to take place. Contemporary culture developed in an industrial society that was labor intensive, but already that culture is changing to one based on knowledge, information and communication, and this demands a change to a service intensive economy that will be the chief characteristic of the post-industrial society.

This is all wonderful and convincing. But at the same time I am faced with the old question which Bell never systematically takes a pace back from the detail of his material to ask - namely: can the creature sustain all this complexity - these numbers, this intricacy of organization? Can this kind of creature sustain it? For there is a hidden assumption in the work of Bell and all like him - that the creature can, in effect, do anything. The limitations of the creature, therefore, do not need to be taken into account. Indeed, they do not exist. Paul Goodman in 1960 castigated this kind of thinking as "the final result of the recent social-scientific attitude that culture is added onto a featureless animal, rather than being the invention-and-discovery of human powers" (*Growing Up Absurd*). At about the same time Clifford Geertz had the same insight:

> Man is to be defined neither by his innate capacities alone, as the Enlightenment sought to do, nor by his actual behaviors alone, as much of contemporary social science seeks to do, but rather by the link between them, by the way in which the first is transformed into the second, his generic potentialities focussed into his specific performances.
> *The Interpretation of Cultures* (p. 52)

I am going to argue that Bell doesn't really believe his own ideology -that there is a built-in ambivalence; that like me he in fact accepts that "capitalist/industrial" society is only made possible by the grasping hand, binocular vision and hand-eye coordination shaped by 70 million years in the trees; by the solidly planted foot and muscular power developed by at least 6 million years on the savannas; by the "storage and retrieval" and "planning and foresight" capacities of the neo-cortex developed by systematic hunting. In saying that we are shifting from a labor-intensive to a service-intensive system - or whatever, we are saying that we are shifting from emphasis on one part of the human primate repertoire to another. But the brain, with its mechanical adjuncts (to which we are turning more and more desperately) is still the primate brain re-tooled by predation. And it is not an organ of cool rationality: it is a surging field of electro-chemical activity replete with emotion and geared for a particular range of adaptive responses. Force it to try to work outside that range for long enough, and it will react - it will rebel. It will regress to those pristine behaviors (including the very necessary aggressive ones) surrounding its primary functions, survival and reproduction.

The sociologists - including Bell when he is on his guard - write as though none of this matters because the institutions themselves are what are at issue, and they are somehow "autonomous" - detached from the grasping hand, the striding walk, and the self-deceiving brain. All this "organic" stuff can be left out of the equations. Thus giant bureaucracy will fail, perhaps, because of internal structural conflicts or conflicts with other institutions - not because this animal - this *Homo sapiens* - will only put up with it for so long before it subordinates it in some ways: sabotage, "inefficiency," absenteeism, alcoholism, "work-related stress diseases," nepotism, "issueless" strikes, burn-out and straight rebellion. Why, if indeed we can "add culture to a featureless organism," shouldn't we be able simply to "train" people to enjoy large-scale bureaucracy, to "adjust their work-habits and attitudes"? Shortly before he died so tragically early, Victor Turner, the arch-priest of "symbolic" social science, wrote poignantly, in *Zygon*:

> The present essay is for me one of the most difficult I have ever attempted. This is because I am having to submit to question some of the axioms anthropologists of my generation - and several subsequent generations - were taught

to hallow. These axioms express the belief that all human behavior is the result of social conditioning. Clearly a very great deal of it is, but gradually it has been borne home to me that *there are inherent resistances to conditioning.* (my italics)

There are flickers of understanding of this in Bell when he lets his guard drop. For example, on page 163 (1976 edition) he discusses why things have "gone awry" with sociological predictions:

> The first has been the persistent strength of what Max Weber called "segregated status groups" - race, ethnic, linguistic, religious -whose loyalties, ties and emotional identifications have been more powerful and compelling than class at most times, and whose own divisions have overridden class lines.

It is not discussed further, but one immediately asks, "Why?" Race, country (region), language, religion: why do these have their "persistent strength" that is so annoying to predictions stemming from the autonomy of institutions? I think I know. Bell lets it pass. But I'll make my own prediction: their "persistent strength" will make a merry hash of the "post-industrial society" too.

On the issue of the failure of "rationality" in "industrial society," hear Bell:

> Traditional elements remain. Work groups intervene to impose their own rhythms and "bogeys" (or output restrictions) when they can. Waste runs high. Particularism and politics abound. These soften the unrelenting quality of industrial life. Yet the essential, technical features remain.

Where do "their own rhythms" come from, that they are so intrusive? And by what standards do we judge that "the quality of industrial life" is "unrelenting" and needs to be "softened"? Again, I think I know, but Bell just slips this one in with "traditional elements remain." But why do they remain? Why, as we have asked, if we can do anything, don't we just re-train people to find industry or bureaucracy satisfying? Perhaps we are just not very efficient or not ruthless enough (Skinner would say so.) Perhaps. Again, on the literal "inhumanity" of bureaucracy (a central theme in Tiger and Fox, *The Imperial Animal*) here is a remarkable statement by Bell (p. 119):

> In the broadest sense, the most besetting dilemma confronting all modern society is bureaucratization, or the "rule of rules." Historically, bureaucratization was in part an advance of freedom. Against the arbitrary and capricious power, say, of a foreman, the adoption of impersonal rules was a guarantee of rights. But when an entire world becomes impersonal, and bureaucratic organizations are run by mechanical rules (and often for the benefit and convenience of the bureaucratic staff), then inevitably *the principle has swung too far*. (My emphasis)

"Has swung too far"? From what? We are not told. Again, if it's efficient, why doesn't it just work? Why do intrusive elements "persist"; how do we know it has "swung too far"? I think Bell is unquestionably right - more right in fact when he exercises his sound human intuition than when he essays his complex analyses. "Irrational" elements persist. We have gone too far.

If we look at "traditional societies" - those simple tribal and village people that are the stuff of anthropology - there is something quite striking that sociologists in their obsession with the torments of industrial society miss: they could go on forever. They are Lévi-Strauss's "cool" societies as opposed to the "hot" societies which have entered the stream of historical change. If we look at the tribes, the bands, the small villages, it is obvious that they have not only existed from time immemorial, but could go on in the same way forever if they were not hit from the outside. Colin Turnbull's pygmies - existing as humanity in its most "stripped-down" form - could clearly persist indefinitely if they were not dragged out of their forest, put in "progressive" agricultural camps, and allowed to die off from the heat and epidemics.

It seems simple minded to say that they do not have the problems of our society - either the real problems of "adjustment" or the intellectual problems of explaining it. But the annoying "irrationalities" that "persist" -these "traditional elements" - and play havoc with our increasingly designed societies, are their "irrationalities," their elements: tribal loyalty identified with language, physical character, symbols and beliefs. The "natural rhythms of the work group" are their rhythms - the whole social entity is a "work group" with its internal reciprocities and rhythms. And if not disrupted, the "traditional" systems work. They just go on. It is not that they never change, but rather that change is not of their essence as it is with societies in the stream of history. As Tönnies was not afraid to conclude, *Gemeinschaft* is always more durable than *Gesellschaft*. Its lifestyle may not be wildly exciting compared with computer wars and disco bars (I'm not so sure) but it seems satisfying to the members, and it is not plagued with the problems that Bell needs so many pages to analyze.

These small-scale societies have their conflicts, of course. I am not saying that they are all peace and harmony. Conflict is part of life - in fact in trying to get rid of it we may do great damage. But that is not the point. The point is that they work, and they work in my estimation because their very scale is compatible with our "environment of evolutionary adaptation" - the social environment for which we evolved and in which were are *supposed* to exist. Bell's "intrusive" elements in the rapidly changing, ever rationalizing, increasingly bureaucratized societies, are precisely the elements that make small-scale, unhistorical, relatively unchanging societies work. When we have "swung too far," it is from this central point.

The Pit and the Pendulum

But what an abhorrent notion to the "autonomists" and the behaviorists. Since we can do anything, we can change society effectively as long as we can put our "expert" finger on just how to do it. But "it" remains strangely elusive. I have another vision of what has happened in "history." The successive "stages" of "progress" are simply different experiments in departing from the basic pattern (as outlined, e.g., in chapter 6 of my *Red Lamp of Incest*). They are not in fact "progressive states" at all; it is simply that changes in technology allow us to try different experiments. The knowledge and technics used in trying these are cumulative, of course. But this only means that we can attempt even more bizarre "designs for living" than previously. But they never work, in the way that the "primitive" total societies work, because they can never be total societies: societies that tap the whole range of human needs and satisfactions for each and every member. These they fragment, and most so-called "social pathologies" are desperate attempts to heal the fragmentation or, if you like, the alienation and anomie.

In changing into Bell's "post-industrial society," we are simply crawling to another part of the fly-paper of history. We won't escape, and it won't work. But I shall repeat my solid sociological prediction: the "traditional/irrational" elements of the basic pattern of our social environment of evolutionary adaptation will persist and disrupt the new order. Orwell envisaged a "total" totalitarianism as the only final (if not happy) answer to the agony of constant change. But Orwell's super bureaucratic state wouldn't work either, as David Ehrenfeld (*The Arrogance of Humanism*) - his great admirer - so forcefully points out. The "inner party" themselves could never sustain their role. Hope did not lie, as Winston Smith thought, with the proles, but with the fallibility of the managers.

This view of history is close in some ways to the sociologies of change - deriving from certain philosophies of history - that see it as oscillating rather than progressing. (This is sometimes called the "cyclical" view of history, but I think oscillation - the swing of Bell's mysterious pendulum -is more adequately descriptive.) In the grand scheme of Sorokin (*Social and Cultural Dynamics*) - cultures swing between two extreme commitments to conceptions of truth: the truth of the senses and the truth of faith. These are the "sensate" and the "ideational" extremes. Since each is only a "partial" version of truth, overwhelming commitment to either won't work, and the pendulum swings back the other way. The center of balance is "idealistic" - those brief periods of history when the senses are idealized: the body in 5th century Greek art, for example, and in the cathedral sculpture of the thirteenth century. It is a convincing scheme as far as it goes, but it only goes as far as "theories of truth" in explaining the swing of the social pendulum.

If we take my "scale of evolutionary adaptation" as the central point of rest, however, so much more follows. Swings of the pendulum can go in various extreme directions from this state, theories of truth being only one element. I would agree with Sorokin that in the ideal state, "truth" would be a judicious mixture of hard practical wisdom and an acceptance of the "sacredness" (i.e., unquestionability) of certain key values. But there is so much more in the sheer content of social relations that can be derived directly from our knowledge of our evolutionary history of adaptation without hanging it all on abstract notions of truth. I confess I cannot - no one can - at this point detail all the content of the steady, central state. But that knowledge will come (although probably too late). In principle, however, it is only the properties of the total/small-scale society, with the added knowledge of its physiological and psychological underpinnings that modern science is revealing.

In this respect John Friedman (*The Good Society*) and other anarchists and quasi-anarchists are interesting. For Friedman society oscillates between extremes of Individualism and Collectivism, with the "Good Society" of Communalism lying at a kind of center. The communal society, however, can only ever be a kind of guerilla operation inside the extremist lunacies of the other types. But this theory is remarkably like Sorokin's in the sense that each of the super-types tries to push to its extreme and breaks down. The Communal Society is a network of groups, each group having "more than three, less than twelve" members, with "8 +/- 1" as the mode. This is about the average number of adult males in the palaeolithic hunting group. Add twice that number of child-bearing females and about twenty juveniles and infants (a typical demographic structure of a band) and you have the roughly 40 persons of the organic group that we as a species evolved in. (It was in contact with many other similar groups, of course.) I stress organic because Friedman's group is not: it is a parasitical ideological group.

Perhaps what we really need in order to recreate the "communal good society" are organic extended kinship groups. But the bureaucracy hates them, which is why it enforces monogamy, restricts inheritance, encourages celibacy and persecutes Mormons. Be this as it may, some theory of the basic nature of the organic group cannot be avoided.

And this brings me back to Bell and his ambivalence. Despite a theory of change based on "institutional contradictions" - the "disjunction of culture and social structure," he is constantly forced back on the human creature and those limitations that Turner had "borne home" to him. Bell quotes with approval (p. 455) Rousseau's comment that the "universal desire for reputation, honors and preferences, which devours us all ... stimulates and multiplies passions ..." and so on. "Vanity - or ego - can never be erased" Bell avers; "... one of the deepest human impulses is to *sanctify* their institutions and beliefs ..." (p. 480, his emphasis); "... what does not vanish is the duplex nature of man himself - the murderous aggression, from primal impulse, to tear apart and destroy; and the search for order, in art and life, as the bending of will to harmonious shape."

Strong stuff! No. They do not vanish. And if in fact we take and put together all the social, emotional and cognitive items that Bell finds so pervasive yet so destructive of the autonomy of institutions, we would come up with a pretty fair rendering of the "basic" human culture from which the successively more "rational" permutations are disastrous departures.

So I return to my point that these "progressive" changes are illusory: they are merely oscillations about a point - swings of the pendulum further and further away from that naggingly persistent, irrational, but totally human central condition or basic state that is the community fitted to our environment of evolutionary adaptation.

The Decline of the West

Where, in time, is this basic state to be found? The answer is straightforward: in the late palaeolithic, some 15 to 40 thousand years ago. It is really that simple. We were fully-formed modern *Homo sapiens*; we had reached the top of the food chain - we were doing quite a bit better than the other carnivores. Then, with a frightening rapidity, it all began to go wrong - or to go "too far," as Bell would have it. Population was squeezed into the Middle East and southwestern Europe by the ice, and the unprecedented social density thus created led to a burst of self-conscious activity evidenced by the fantastic art of the period (Bell's "search for order in art"?). Hot on the heels of this came the warm interglacial in which we are still living (and which has almost run its course), and the first of the violent oscillations happened - the domestication of plants and animals. After that, the swings of the pendulum went on, sometimes at a leisurely pace, sometimes wildly. At the points between the wildest swings, we get the most terrible upheavals and carnage; and each huge swing has the effect of sending the pendulum wildly off in another direction. The only "progress" in

These swings are roughly illustrated in the diagram. The "upward" movement is merely chronological and does not imply progress, except as cumulative technological change. Also the extremes of the swings are simply my own highly condensed judgments; other observers will stress different ways in which the swings went (or will go) too far. For example, under blanket headings like Feudalism and Industrialism all the effects of these new systems of production have to be included. Thus I have not included Industrial Capitalism or Socialism or Welfare Liberalism etc. under Industrialism, since these are all effects of the industrial evolution. Nor have I added refinements like Monopoly Capitalism or Multinational Corporations or Imperialism since these again are subdivisions of the more general headings. Again, the major wars I have indicated are those that have reflected or led to features of the major shifts. Thus the Fall of Rome and the Barbarian Invasions led to Feudalism; the Franco-Prussian War (see Michael Howard's excellent description of it) and the American Civil War reflect the impact of Steam Power which led to railways, and the "nation in arms": universal conscription in France; the dominance of Ideology in America. These were the first great modern wars.

It should also be remembered that many cultures did not participate in these shifts, at least until the Western powers forced the results onto them. It can be argued that the picture is Euro-centered. True. We are not here presenting a scheme of World History, but a map of the major swings of the "progressive" pendulum, and these mostly took place in Europe after the Middle Ages; a fact that has obsessed modern social science. We cannot help but be centered on the West, and in consequence on the decline of the West, for this is where the pendulum did its latest and most damaging swinging. Bell is only pointing the way to another such swing, and at the same time realizing that something human is here being denied. I guess that all I want to do is keep calling attention to this humanity and to plead against its denial. If we can't go back to the "palaeoterrific" then perhaps we can at least drop the nonsense about progress and rationality and start thinking about how we can serve that stubborn human core within the context of the inhuman super society. Perhaps it is only marginally possible. But it certainly won't be possible at all if we don't recognize the problems.

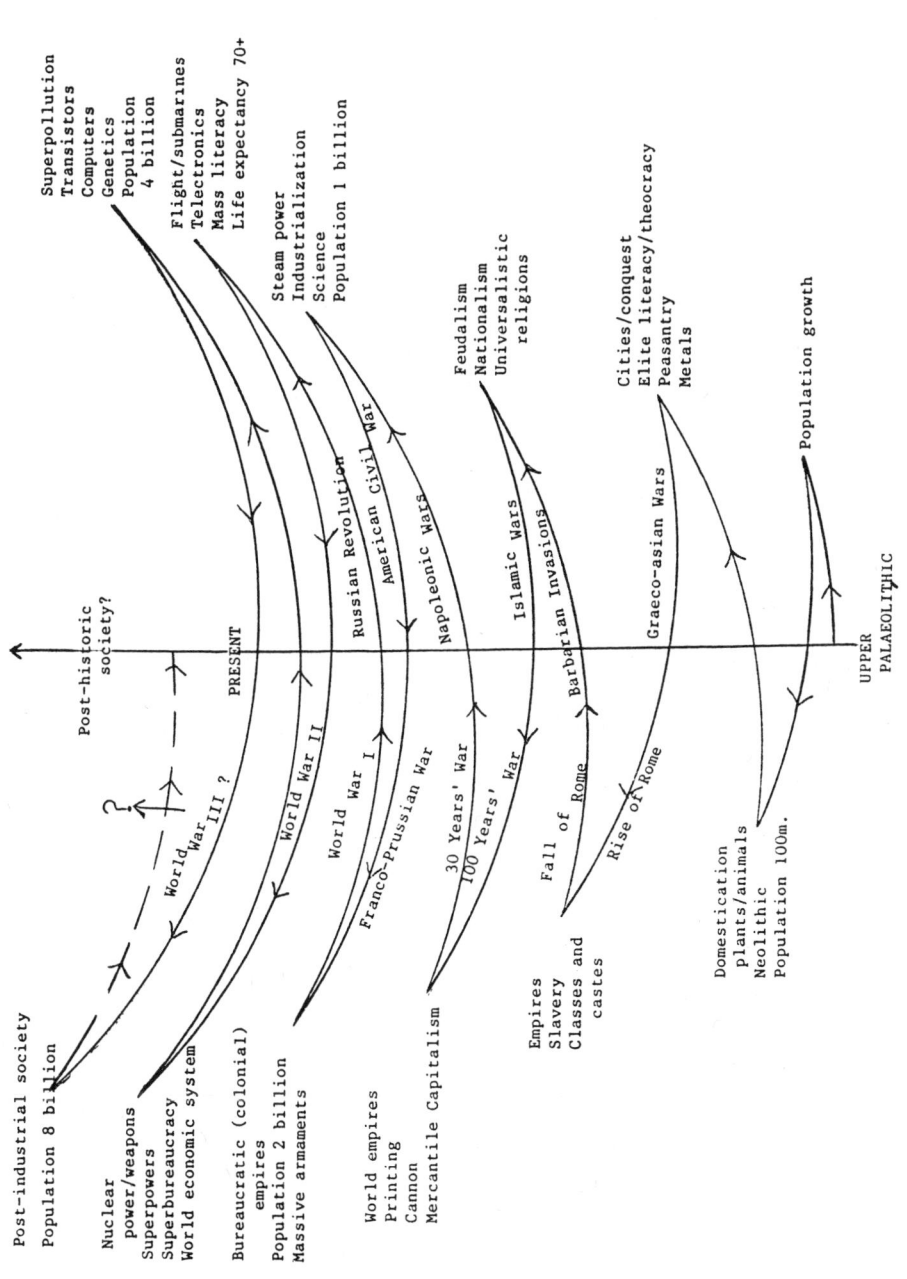

As Others See Us
Doris Lessing says, in *The Sirian Experiments*:

> I think it is likely that our view of ourselves as a species on this planet now is inaccurate, and will strike those who come after us as inadequate as the world view of, let's say, the inhabitants of New Guinea seems to us. That our current view of ourselves as a species is wrong. That we know very little about what is going on.

It would be hard to press this view, I suppose, on all the Daniel Bell's of the world who have written at such expert length about what is going on, has gone on, and is going to go on. That is what makes it hard for me to be a social scientist, for I agree with Lessing: our current view of ourselves as a species is wrong; we know very little about what is going on. But at least since Darwin we know a little more about ourselves as a species, and in another place I will have to analyze why the social sciences have spearheaded the refusal to acknowledge even that. But consider - what possible headway could the view that all "history" is a series of wilder and wilder divergences from a palaeolithic "norm" make today in the world of behavioral science orthodoxy?

I am always amused at the "Marxist" objection that the biological approach to social behavior, as they love to say, "de-historicizes" man! What a puffed-up, pre-Darwinian, self-satisfied Enlightenment view of themselves these *soi disant* "Marxists" have! It is not biology that "de-historicizes" man but a view of him that would deny the relevance of *millions of years* of his history! What they call "history" is a problematical blip at the end of this trajectory - the wild swings of the pendulum shown on the chart, swings that represent only a fraction of one percent of his history.

One part of the view of ourselves as a species that is wholly wrong, then, is the conventional time frame in which we choose to analyze ourselves: as though the last few thousand years are peculiarly privileged and the rest can simply be written off as "prehistory"! This is pride, arrogance, hubris of a high order, and we are paying for it. Before Darwin, before we knew, it was perhaps forgivable. Now it is not even funny. It is the root of our self-destruction. Kenneth Bock can castigate "sociobiology" for not being able to "explain cultural diversity or recent history" without, apparently, any insight into the irrelevance of this claim. It is a measure of our lack of vision that an overwhelming majority of social scientists would agree with him. We have seen Bell's "explanation" of "recent history." As an anthropologist I have spent my life "explaining" cultural diversity. Blips on the end of the trajectory. In a thousand years those successors Lessing envisages will look back (if there are any successors, of course), with some passing interest, on the phase called "history" by those chrono-centric myopians who lived through it. The successors will be mildly amused by their arrogant pomposity but, we hope, will have learned never to repeat it.

Aside - For the Social Scientist

Let us, as a late and unlamented President said, get this perfectly clear. The upper-palaeolithic was in balance between the organism, the social system, and the environment. It got pushed off balance, probably by a sudden (in relative terms) increase of population density in southwestern Europe, the Middle East and parts of Asia. Once the consequences of this density began to take hold, there emerged certain social properties that had not existed hitherto. And Durkheim was right that these were social properties, not individual. He was wrong, however, to divorce them from the biology of the organisms, for only this made them possible.

Thus, for example, religion in a small tribelet was one thing, but became something quite other when large numbers were involved with the concomitants of classes, religious castes, power elites, exploitation, etc., etc. The raw material was the religiosity of the palaeolithic hunter: the "new look" was the emergent properties of the new social density. Ever since, these emergent properties - seized as their subject matter by the social scientists - have been on a collision course with the social needs of the palaeolithic hunter. This is what we call "history"; and "cultural diversity" and "recent history" are both but the latest set of examples of the collision course (for better or for worse, since some versions of culture are closer to the balance).

So to say that "evolutionary biology" cannot "explain" recent history is a totally empty claim. What could it mean? That it cannot account for specific series of events in recent time? Of course it can't. What can? This view entails a curiously nineteenth-century mechanistic version of "explanation." But if what we are after is understanding or perhaps interpretation, then only evolutionary biology can help us. The historians and social scientists are too trapped in the myopia of their short-term vision to help us. They have taken the emergent properties of the post-palaeolithic society as a reality *sui generis*. But they do not understand the genesis of this reality and therefore its ultimately fragile and dependent nature. For it is dependent on the continued acquiescence of the palaeolithic organism, and this, as Bell sees (but does not pursue his vision) it does not have. I would "explain" recent history as an ever more desperate attempt to marry this precarious acquiescence to an ever more runaway technology. And that's from evolutionary biology.

If I can put this in a succinct phrase for the Bocks of this world to ponder: It is not so much that evolutionary biology explains what we do, as that *it explains what we do at our peril.*

I don't want to trespass on Lessing territory, since while I certainly think that the science fiction writers often have a better grasp of things than the social scientists (Frank Herbert, for example) I don't aspire to their version of the vision. I am happier defending my own insistence on the virtues of the palaeoterrific. (I don't know who coined the term - and I think it was meant sarcastically. But I don't care. I'll just reverse the sarcasm and shamelessly appropriate the lovely word.)

Et in Arcadia Ego
There is a reflex tendency to dismiss this view of the palaeoterrific as romantic primitivism. But for me it is based on a hard-science view of the evolution of human behavior. It also gets around a common criticism: that those philosophers who advocate "living in accordance with nature" have a hard time defining that

very "nature" they are so fond of. (Thus for Diogenes it was living in a tub; for Aristotle it was living in a city.) I do not have that difficulty: my "nature" is located during a specific era of evolution for specific reasons advanced above. It is not a vague conceptual "state" defined by whatever the philosopher feels is "natural" and therefore by implication "good." It is a specific time and even place; and it is not "good" because it embodies qualities I or any other philosopher admires necessarily. It was not, probably, according to most definitions so far advanced, very "good" at all. Life in it was not necessarily always nasty and brutish, but it surely was some of the time; and it was certainly short, by our standards, for most people. I would plead for a moratorium on the use of "good" here, since it is very confusing. I am going to stress over and over again that I do not think we should hold to the palaeolithic model because it is some kind of utopian ideal, but because it is what we are and it worked because it stayed within those bounds. And that is all. Our notions of "the good life" are now so corrupted by all the paraphernalia of literate civilization that we have probably no way in which even to talk about the virtues of adaptation *per se*. That is, we cannot even conceive of approving a way of life simply because it worked, in turn because it was the way of life to which the creature had evolved. We see this for animals when we debate about keeping them in zoos which we "reform" to approach the untouched grandeur of the natural surroundings of the incarcerated creatures. What I am asking is for a similar approach to ourselves. We have left that untouched grandeur and the life associated with it and have entered what Morris chose to call *The Human Zoo*: civilization. Can we, or can we not, reform that zoo so that it at least approximates the conditions from which the unfortunate animals were removed, or in this case, removed themselves? This is a crucial point in which we differ from the other animals, and I must return to it.

But before I do, let us consider the issue of the virtues of the palaeoterrific. I will stick to my point that I am not looking at it as utopia or dystopia but simply as palaeotopia: as what it was. But I still feel inclined to ask: would life in the upper palaeolithic have been so unsatisfactory for those of us nurtured in "civilization"? We tend to view this from our softened position as inheritors of a gadget-ridden, mass-literate, high-technology culture. Let me do a little thought experiment here - a highly personal one, as a thought experiment has to be - and project myself back into that protoculture.

When I view my own life, I find that many of the things that satisfy me most are not dependent on the high-technology culture, nor are they the things for which I receive reward and praise from that culture. On the other hand, some things for which I am rewarded have their palaeolithic counterparts. Thus sitting on committees is equivalent to taking part in the councils of the tribe; teaching the young is part of the system of tribal initiation; and tending the myths and legends of the tribe (social theory) is pure palaeolithic - even if it then was solely oral rather than semi-literate. Literacy is a late intrusion which we have not properly understood or assimilated, which has some very bad side-effects, and which is probably much overvalued. Most people get by without it, and many of those nominally "literate" manage with a minimum and revert to orality (television) out of preference.

When I was young and poor I used to borrow books from the library, and, so that I would always have them to hand, committed to memory large parts of Pope's Homer, of The Idylls of the King, the Lays of Ancient Rome, The Pied Piper of Hamelin, etc., etc. I can remember accurately the whole of many operas (particularly Gilbert and Sullivan), oratorios and cantatas, masses of Shakespeare, and much, much more. And I am here talking only of accurate memory, and only of a small part of it. My memory storage is nowhere near exhausted, and a whole mass of memorized songs (running into four figures) were learned orally. Given living poets and musicians rather than books and scores, I would have perhaps learned even more, and learned it more quickly. Literacy slows learning if anything. The Greek poets recited all of Homer by heart; the prodigies of memorization of the Celtic bards is legendary, and is matched today by the reciters of folk epics in the Balkans; read Mark Twain on the memories of Mississippi steamboat pilots; look at the inventory of a Navaho singer. Mechanical devices make us lazy. My point is that I would have enjoyed a good intellectual existence in an oral culture without much loss, among people similarly well-stacked with information, and even some division of intellectual labor. It is one of the things that can make growing old in such a culture so superior to growing old in our own: the old are the living libraries, the *bioblioteques nationales* of the tribe - and are often cherished accordingly.

I would have been valued for the lullabies I composed to quiet the children, and the songs I made up to amuse them. As to philosophizing, anthropologists now know that the myths and rituals of a tribe are just that. Fine tuning of old myths to new social and cultural needs is what any ideological system - like "social theory" - is about. I would have been good at that, along with tribesman Bell. I've always preferred couching ideas in story or verse rather than in "rational" argument, anyway. (Another late and still unproven innovation.) I "dropped" music, poetry and art at which I was also amateurishly good - to take up "science," thus losing the unity of art and science that is inherent in the logic of myth and ritual. How typical of our arrogance that we now honor a person who is good in many "fields" - arts and sciences, sports and technologies - as a "Renaissance Man"; we never think to call him a "Palaeolithic Man"! But then, we haven't even absorbed the astonishment of Altamira and Lascaux.

I could have been on the "committee" that organized the incredible Lascaux ceremonial complex with its great halls, tunnels and side chapels, its crypts and vestries. (See John Pfeiffer's *The Creative Explosion.*) If we look later, as Henry Adams did to his central point of human unity, the eleventh, twelfth and thirteenth centuries, what do we find? We find again great ceremonial centers organized by committees at Chartres or Paris. And what are they but huge, dimly lighted caves erected above ground, with crypts, tunnels, side chapels and wall decorations; erected at huge expense of money, time and labor, and more often than not to honor the Great Mother - Notre Dame - who is Nature. Adams - who never really grasped Darwin, having been sidetracked by Lyell - got as far as the eleventh century in his search for the time of unity to compare with the time of fragmentation he was living in. He was on the right track, but he didn't know about Lascaux and Altamira.

Insofar as these were ceremonial centers (like later Stonehenge) where many different people came for initiation into the mysteries, I would have been good at picking up the foreign tongues and appreciating the power of the magic of alien tribes. When it came to the hunt, I think I would have managed once I knew that the ferocity of wild beasts could be countered by the collective power of human imaginations and intellects. I found this out directly when facing an enraged and dangerous wild animal in an arena, armed only with these skills and a piece of cloth. As to war - or such skirmishing as passed for war then - well, I am the son of a soldier, and I would have had something worth fighting and dying for: my genetic investment in the little tribe. Men do not really, except in heroic epics, die for the ashes of their fathers and the temples of their gods, but for their living family of fellow creatures - as Macaulay, to his credit, has Horatius go on to recognize. They would really have needed me, in a way I find hard to believe that the "Free World" needs me, or the United States, unless it is disguised as an "uncle" to tap my palaeolithic kinship guilts. No. I don't think the palaeolithic would have been all that bad at all, and I've nowhere finished the list. It would take a book to spell it out. It may well have been a short life, although it would have been free from epidemic, stress, and pollution-related diseases; but better that filled with immediate human and intelligible meaning than a lingering fragmented existence, with less and less sense of worthwhileness as the species connives greedily in its own extinction.

Friends who have read the foregoing and who know my perhaps over-civilized tastes maintain that I must have my tongue in my cheek. Of course, this has been a "thought experiment." Of course, there would be much I would miss if thrust back now into a palaeoterrific existence. But again, of course, I am not suggesting some such reversal - a sudden taking away of all we know and a thrusting of us back into the caves. I am assuming the position of someone like myself (a similar personality) who knows nothing other than the palaeolithic. What I am trying to convey is that such a person (and I take myself as an example precisely because I am overcivilized) could have lived a rich, full, meaningful, satisfying, exciting and no less, perhaps even more HUMAN life in this period. That is all. I am trying to object at the personal level to the notion of "progress": that because we are now surrounded by and totally accustomed to the paraphernalia of industrial civilization, that to be stripped of this would reduce us to the level of "degraded savagery" or "the stone age" or "brutal barbarism" or something such.

We still automatically think in this way - this eighteenth/nineteenth century social evolutionary way, and no amount of enlightened social relativism cures us of it. (Although Jean Auel's *Earth's Children* series, and her charming creation, Ayla, seem to have gathered quite a following -but I suspect this is largely among environmentalists who mistakenly see palaeolithic man as the original conservationist, rather than among anthropologists who tend to be scornful of this, in fact quite interesting, exercise in conjectural history.) Indeed, my cultural relativist anthropological colleagues are among the worst offenders. The notion that we have not only technologically but morally progressed since "the caves" is deeply ingrained: "If it weren't for X, Y or Z we'd be back in the caves." And even if not morally then at least, they insist, intellectually. You could not really contemplate, they say, going back to a pre-literate life - you're the most avaricious reader we know! And to abandon not only the literacy, but all that sheer knowledge - the rational scientific knowledge - and sink back into "superstition" etc., etc. It is no good pointing out to them again that this belies their relativistic premises. If the "little Society" or the "savage mind" of which they write so eloquently and which they defend so passionately is as god-damned wonderful as they claim, then why be afraid of being part of it?

Is it not they who are pretending in fact? What it boils down to - what they really do not want to lose - are sheer material comforts that enable them to extol the virtues of the primitive state while keeping it at a vast distance with their gadgets and life-prolonging technologies. As a palaeolithic hunter-shaman-warrior I would know nothing of these things and hence could not miss them. Would I be less or more "human"? Such a person transferred (like Tarzan or Ishi) to our own times would be suicidally nostalgic for his own way of life. It is all relative. It is nowhere written that it is better in any absolute way to be a Western-Rational-Scientific-Literate person than to be a palaeolithic Hunter-Shaman-Warrior-Artist-Poet (oral). I am stuck with being the former, but I recognize this as an aberration, not a result of "progress" - just a huge deviation.

I shall struggle, reproduce and die just like my counterpart. I have a different set of artifacts to do it with, that is all. My word processor has not composed the Iliad. My acrylics have not painted the great bulls of Lascaux, and my camera only obediently reproduces them. My contemporaries are diverting the energy of the human population into the means of its own destruction. I have books and central air-conditioning and jet travel and a VCR. Medical science may well see to it that I and billions of others enjoy these until we are eighty or more. Wonderful.

For those more interested in the logic of this argument than its charm, I should emphasize again that I do not need this "virtues of the palaeoterrific" premiss. I only need that the palaeolithic (upper) is our Environment of Evolutionary Adaptation (EEA). I just think it should get a better press, that's all. A more serious criticism would be that even by the late palaeolithic we had passed the point of no return - with, for example, the invention of the bow and the destruction of the megafauna in Europe and North America. It could be argued that the EEA should be set earlier. But I would hate to give up the cave art and all it implies. I admit this as a weakness, and one that, if basically human and hence generalizable to our ancestors, may help explain the inevitable start-up of the pendulum swings.

The Reach Exceeds the Grasp

Which leads to the inevitable question: could we have stopped there? I very much doubt it. And that is the human tragedy and its grand paradox. I once asked an engineer why cars were built to achieve speeds (like 120 m.p.h.) that they almost never attained and couldn't use except for brief periods. He explained that if you wanted a car to have a cruising speed of, say, 85 m.p.h., then it had to have a capacity of 120 m.p.h. - to maintain the eighty-five. The problem is, with an ambitious driver, there is a temptation to want to cruise at higher and higher speeds.

Could it be that the palaeolithic brain (which is our brain, it hasn't changed) was geared to cruise at the palaeoterrific level, but in consequence had to be capable of the industrial and post-industrial societies? That what we have done, post-Altamira, is put our foot (as it were) ever more firmly on the cerebral gas pedal simply because we *could* go faster? And that now we are trying to cruise at what were only intended as passing or emergency speeds? We are doing 110 in a car designed for 85, because we can do 120. But, I asked the engineer, for how long? Not very long, he said. It's all relative of course, but not very long.

We have never recovered the balance, never found (except in odd corners here and there) the cruising speed. We are roaring out of control - wild swings of the pendulum - but for how long? Can we retain the technology, but regain the cultural and social center? We scorn as "savages" and "backward" or "under developed" those who kept close to the balance, and we force the technology on them both to salve our guilt and extend our markets, and because we have persuaded them that they are chicken to be going at 65 when they can do 120. In their cruising state they are not necessarily "noble" - but perhaps they are closer to being "human" in a sense we may never again understand.

The anthropologists have partly understood it - but they are viewing it from the speeding car. Their applause, reward and symbolic immortality is located at one end of one of the wild swings of the pendulum - the extreme of rational, bureaucratic science; they cannot conceive of a return to the center, even if they see its virtues better than most. And as we keep observing, perhaps there is no return. But except in technology I do not see the "progress." And even that is not clear since the technology is destroying us; or we are destroying ourselves with the technology; or whatever.

In some sense then, we may be trapped. We - the creature - produce these fantasy structures - cultures, religions, laws, civilizations. We produce them out of the raw material of our speeding brains and ricocheting imaginations. So they correspond to something in the creature or they would not work at all. Humans are eternal adolescents: put them in a fast car and they'll speed. The raw materials of testosterone and curiosity will see to that. And they do correspond to something: one dimension of our humanity is ingenuity and deceit. Capitalism saved itself (for a while) by appealing to the greed of its workers. "Affluence" bought them off. But the rest of the world, it appears, cannot be so easily bought and stagflation threatens capitalism again. Even the socialist world, however, seems glumly to accept that sheer repression cannot work and that it must appease the greed of the palaeolithic glutton. In the West we are back to bread and circuses. (Which only fail when the bread is mouldy and the circuses boring.) Television for all with electronic games for the children (with a return to low-key materialism) will buy them off -stem the dissatisfaction for a while.

But all these things appeal ever only to part of the creature's emotional and intellectual needs. We are greedy, certainly, but not endlessly greedy. We also have pride and a craving for attention for example. Media society can buy us off here too, for a while. As Andy Warhol summed it up, this is a society in which everyone can be famous for fifteen minutes. But the nostalgia for wartime that continues to embarrass pacifists reflects a need to feel integrated into the group when it faces danger. (Now where could that need have arisen I wonder?) "We were all together then. People helped each other. We weren't going to let the bloody Huns get us down." This refrain as part of the very real nostalgia among civilians suggests the deep needs involved. Bureaucracy cannot provide for them. But the sociologists will not face these issues, except to take a passing glance and hurry back to the autonomy of institutions. This is probably why the sociologists, like the economists, are constantly being faced with mass phenomena they can't predict, can't explain (except with after-the-fact rationalizations), and wish would go away.

Words of Our Fathers

Actually mine is not such an outrageous social science notion as it may seem. I am only taking to a logical conclusion, and in the light of more recent information, the "loss of community" theme running through much of modern Western social thought. It could perhaps be said to be *the* theme of modern sociology. But with the possible exception of Tönnies (and Kurt Vonnegut) few thinkers have wanted to push it this far; that is, to claim not just that we are losing - or have long ago lost - a more "sociable" form of existence, but that we have lost this because we have lost our setting in nature. We have stepped irrevocably outside the limits of our environment of evolutionary adaptation. Even the anthropologists (like Redfield) saw the "little Society" as a cultural artefact, not as an "organ" adapted to its evolutionary niche. Max Weber, however, in his admiration of - and projected book about - Tolstoy, came close to this point. This is the extra distance I am prepared to go. It is only a way of translating into the language of social theory the not uncommon complaint that the modern world is out of step with "human nature."

It fascinates me that one of the most perceptive social analysts before Bell should have come close to this same conclusion. And Karl Polanyi, in 1944, was also writing of a major shift in the swing of the pendulum - precisely the shift from the pre-industrial to the industrial-capitalist society (from which, if Bell is correct, we are now lurching into the post-industrial phase). Polanyi called this *The Great Transformation*, but throughout warned us that the transformation itself has muddied our thinking (I quote from the Beacon Press edition of 1957):

> The habit of looking at the last ten thousand years as well as at the array of early societies as a mere prelude to the true history of our civilization which started approximately with the publication of the *Wealth of Nations* in 1776, is, to say the least, out of date.
> (p. 45)

This sentence could well have stood as an epigraph to this argument. Except, of course, that I would have wanted to say the last ten million years! But Polanyi had the point. He says, on p. 46:

> If one conclusion stands out more clearly than another from the recent study of early societies it is the changelessness of man as a social being. His natural endowments re-appear with a remarkable constancy in societies of all times and places; and the necessary preconditions of the survival of human society appear to be immutably the same.

Thus it is that "traditional elements remain." Thus it is that the "natural rhythms reassert themselves." And so on. In his great conclusion on the inherent inconsistencies in the "market mentality" Polanyi asserts it once again having brilliantly analyzed the examples (p. 150, my italics):

> For if the market economy was *a threat to the human and natural components of the social fabric*, as we insisted, what else would one expect than an urge on the part of a great variety of people to press for some sort of protection? This was what we found. Also, one would expect this to happen without any theoretical or intellectual preconceptions on their part, and irrespective of their attitudes towards the principles

underlying a market economy. Again this was the case. Moreover, we suggested that comparative history of governments might offer quasi-experimental support of our thesis if particular interests could be shown to be independent of the specific ideologies present in a number of different countries. For this also we could adduce striking evidence.

Thus the "free-market liberals," he points out, were the first to cry out for protection against the consequences of their own theories with the enactment of anti-trade union and anti-trust laws. And thus again and again when the "human and natural components of the social fabric" are threatened, natural and social humans recognize unconsciously that things "have gone too far" and strive for redress even in the teeth of their own theories to the contrary.

A View from the Bridge

But despite such interesting insights - and there are many such in the history of western thought - we are going to have to be ruthless in treating most views of man and society as largely beside the point and interesting mostly as data. They cannot be much use in the construction of a future adequate social theory because they all suffer from the same problem: they operate from inside "history." They have no view of the place of history in the total story of the human species. Anything before history is dismissed as "pre-history" and its only interest is as a backdrop to the historical period. Various "philosophies of history" or works of political or social philosophy may well tell us important truths about this period, but they could not and cannot put the period itself in perspective since they have no idea what the perspective is or even that there is one! The "State of Nature" was dealt with in a few (usually erroneous) sentences. Since Darwin the possibilities of a longer perspective have been there and the "early condition of mankind" got some attention. But even so, until very recently, no one knew the length of time that the so-called "early" period covered. We now know that it was at least five million years.

The Darwinian news that we must have emerged from the animal kingdom was shattering. But until we knew the time period

in a temperate interglacial (which is coming to an end) and represents much less than one percent of real human "history" properly understood. We arrogantly refer to "early man" in a few sentences, when it is ourselves - "late man" - who deserve a passing mention. As we have seen, Lessing prophesies that some "historian" of the future will look back with amusement at the pathetic pretensions of "historic man", much as we now laugh at the superstitions of the savage, the ignorant peasant, or the Middle Ages.

The wisdom of the Greeks, of the Jews and the Christian scholastics, the confident learning of the Renaissance and the Enlightenment, the turbulent self-assertion of the Romantics and the triumphs of Scientific Method and Technology, while full of insights, will have to be put on hold.

The new sense of ourselves as a problematical and experimental blip at the end of the trajectory of human history has to take hold and everything has to be recast in that perspective before we can even start to think sensibly and constructively about ourselves, our behavior, our values, our societies. All our assumptions - all that accumulated detritus on all those library shelves - will have to be re-examined and probably largely discarded. Discarded that is as truth. It is all useful as data on the aberrations of the historical period, and it is always, at least to me, interesting to see how close to a real insight into the human condition many thinkers and poets could come without the benefit of any knowledge other than their intuitive knowledge of their own fragile humanity. Which brings us back to Bell and my contention that he is closer to the truth when using that good intuition than when wielding the dubiously effective club of sociological theory.

The Cool Society

What I see when I read the complexities of Bell's account is Society the Great Leviathan - the stranded beast thrashing and heaving in a desperate attempt to live, as it becomes massively bloated, poisoned and tormented. Bad as its condition is, there is a resilience; there is endogenous healing power; there are social antibodies and a cultural immune system. Leviathan struggles to heal itself; struggles to restore the basic healthy condition it knew before its ambition brought about its near collapse. The healing process is itself painful and the beast suffers much. Now the

before its ambition brought about its near collapse. The healing process is itself painful and the beast suffers much. Now the problem is this: the diagnosticians of the wounded state have mistaken this state for normality. So they see the pains and suffering of the healing process as pathologies! (Such as the "epidemics" of illiteracy, teenage pregnancy, divorce, juvenile delinquency, drug taking, and terrorism.)

The frightening situation therefore reveals itself: they rush to cure what they see as pathologies, but what they are doing is hampering the healing process! Those a little more perceptive say: Leviathan had become used to the wounded state, it wouldn't have killed him, but the healing process is too dramatic; he has to learn to live with the wounds (to "adjust"); they are normality now. The metaphor breaks down because we diagnosticians are part of our own Leviathan. We may even be one of the pathologies. At least this can act as a cautionary tale: the physician may be part of the disease.

This view of the human condition is a version of what I and others have said often enough. The previous paragraphs are quotes, in fact, from a previous book. There also I said that any of these diatribes are only contributions to a larger project, the aim of which is to free us from the intellectual shackles of the Enlightenment faith in Reason, the Romantic passion for the Individual, and the Nineteenth-Century worship of Progress. But it is worth saying over and over again because no one gets it the first time. The prejudices are too ingrained and the urge to kill the bearer of bad news too deep-rooted.

"But we can't go back!" Of course not. And I think like Friedman that we cannot successfully use our new knowledge to "plan perfect communities." Huge equations are involved and there are too many unknowns. In any case, we always have to start from where we are and tinker with that. (I suppose at heart I am a Burkean conservative, and shrink from radical-utopian solutions, including those embraced by the far right.) I was taught by Karl Popper and still believe in "piecemeal social engineering," but with a difference. We must do our patching and mending on some other basis than vague liberal-humanitarian notions of the good. These often result from a merely negative view (oppression is bad etc.), and also often represent quite violent swings of the pendulum

"steady state" of human society (see *The Red Lamp of Incest* chap. 6), and engineer our way as close to it as possible. The classical anarchists - Kropotkin particularly - saw this really quite clearly. They drew this and not the competitive/laissez-faire moral from Darwin! The "small is beautiful" crowd see it too - but they have no basis in evolutionary biology to direct them positively. Small is beautiful, for them, simply because big is bad. But they have stumbled on part of the truth.

Is there any hope of restoring the basic pattern in the mass societies and coercive states of the post-industrial future? The optimist in me says: Yes - if we can hang on and assimilate the new knowledge, and if advanced technology can free us from total preoccupation with work, and if we can reduce and reverse population increase, and if we can obliterate the effects of pollution etc., then we can engineer small-scale polities (under the technological umbrella) that will be real experiments in social living within the basic pattern (not communes!). This is the dream of Lévi-Strauss. Rather engagingly he sees us going back to playing with very complex kinship systems, for example. This would indeed be close to the core of our palaeolithic social being - the assortative mating system!

But the pessimist in me says: No - we cannot stop the trends however pathetically we try to rebel. The pendulum will swing ever more wildly in our efforts to "design" a livable environment for unlivable materials. We will poison ourselves, overcrowd and starve ourselves, or, more probably, blast ourselves out of existence, before we get a chance to try the piecemeal engineering that could lead us, not to utopia, but to a humane scale of social existence (violence and all).

Religio Laici

So far, and leaving socialism on one side for the moment, the ethical guidance we have leaned on most heavily is derived from the great universalistic religions. But the concerns of the founders of these universalistic religions seem strangely out of place in small tribal societies - anachronistic and irrelevant. Those concerns with sin and salvation, guilt and recompense, rejection of the world and self, brotherly love and moralistic neighborliness, self-sacrifice and communalism, temperance and abnegation: they are simply either

self, brotherly love and moralistic neighborliness, self-sacrifice and communalism, temperance and abnegation: they are simply either out of place or totally unoriginal in palaeo-society. But we should note these two possibilities: where they are out of place it is because they are appealing to guilts and problems that scarcely exist in the small-scale society; and where they are unoriginal it is because they are asserting moral platitudes that would be unremarkable in such a society.

What we see in the great world religions then, is a kind of moral desperation at the state of man and the world - which is why they continue to appeal to us today. Faced with fragmented, alienated, self-seeking, individualistic and guilt-ridden "civilized" man, the great teachers -Buddha, Christ, Mohammed, Confucius, St. Francis, Thomas More, Rousseau, Tolstoy, Gandhi - regardless of their particular delusional systems, have issued *clarion calls for a return to palaeo-morality*. They have urged a return to the communal ethic of the tribelet. They have differed about where the boundaries of the tribe should be drawn. Of the great three, only Buddha seems to have been truly universalistic: the whole of mankind was his tribe. But the followers, with their proselytizing zeal, wished all mankind to be included, either voluntarily or if necessary by force. They are sad and moving these desperate attempts to stem the progressive-historic tide. But why they are interesting for us is their key insight into the loss of the palaeo-morality and the urgency with which they used all the metaphysical armory at their disposal to fight for its reinstatement. On some things, for example pacifism, turning the other cheek, and loving one's enemies, they were way out of line with the palaeo-ethic. This is one of the instances where the tribesman would find them incomprehensible. But the pursuit of enemies and the exacting of vengeance are intermittent things with the tribe; they do not claim many victims and they do not threaten the social fabric. In densely populated, stratified, colonial and civilized societies of the Mediterranean, Middle East, India and China, on the other hand, violence was rightly to be feared as a more total threat. The tribesman would not have understood. Mohammed, a tribesman speaking to tribesmen, would be, and has been, much more appealing. His legacy, when multiplied by millions, may well usher in the end, and this would be the ultimate paradox of consciousness out of context.

messages. The followers had no choice. They were caught up in great oscillations of history and could no more resist those inexorable movements than we can for all our technological and rational-scientific sophistication. The followers were realists who knew that they must cooperate with the inevitable, and hence adapt a morality and metaphysics that sought to oppose the inevitable. They held up the assertion of the palaeo-ethic as an ideal, but one which, given the sinfulness of mankind, they had little hope of achieving. Part of their embarrassment stems from the very absoluteness of the founders' ethics: it was all or nothing. And this is perhaps our lesson: we may perhaps hold up the palaeo-morality as an ideal, but we must try to restore such of it as we can in relatively limited circumstances, recognizing that we have long ago passed the point where we could insist on absolute and universal application. That point lay in the very early neolithic when smallness of scale still operated, and still operates in those small-scale neolithic survivals studied by anthropologists.

This is not to say that communalistic ethics are universally followed in small-scale societies. To say so is to make the same error as to say these societies are free from conflict. We are not talking here of little utopias or of noble savages at all. But we are talking of what is accepted in such societies as a moral ideal, either explicitly in codes or implicitly in custom and myth. No society is known which does not have sanctions against wrongdoing. These would scarcely be necessary if there were no wrongdoers. But the important thing is that the very institutional structure, plus the small scale, of these societies, makes dangerous deviation difficult. In trying to restore the palaeo-morality as much as possible then, we should ideally aim, not as the great religious teachers have done, to impose explicit moral codes through supernatural or other sanctions, but to build compliance into the structure of our institutions; a compliance that will be absorbed through socialization and flow from a free acceptance of "our station and its duties" (Bradley). This, rather than either total disorder or total obedience, is the true dream of the Anarchists and the Idealists. It would be a society, in other words, where the rewards for communalism so outweighed those for individual assertion that no reasonable person - and most of us are reasonable most of the time - would even consider the alternative; it would not seem to exist.

reasonable most of the time - would even consider the alternative; it would not seem to exist.

This may be beginning to sound like a heady call for perfection of the kind I claim to eschew. But if the reader is really aware of the argument here, then it should rather appear as heavy realism. We cannot, I have insisted, create moral utopias in our present circumstances. Nor should we try, for such utopias are foreign to our palaeo-nature which is a mixture of good and evil and must allow for the expression of both. We *can* attempt a realistic, piecemeal social engineering of our over-individualized, over-rationalized, over-bureaucratized and altogether too large societies, to produce small-scale institutional settings where the palaeo-morality would have more chance to flourish. Such settings will have conflict; they will have violence; they will have wrongdoing; they will even have wars. But these will all be on a human scale; and while we will not eradicate conflict and evildoing, we can render these manageable in human terms. There may be killings, but there will be no Auschwitz; there may be wars, but there will be no Armageddon; there may be heresies, but there will be no Inquisition; there may be crime, but there will be no Syndicates; there may be repression, but there will be no Gestapo. Now while I say this is hard-headed realism I do not say it can be achieved. I say it may be the best we can hope for, but I am totally pessimistic about its achievement. I doubt if we shall ever recover from our delusional systems (the "disease of rhetoric" as Doris Lessing would have it) long enough ever to consider this alternative.

If we do, however, we must face a common criticism. It has often been observed that collectivism and individualism are in constant conflict, and that while extremes of either are bad, it is hard to strike a balance (Friedman, op.cit). Although the moral problems of rampant individualism may be obvious, the soul-destroying alternatives of total collectivism are by now unthinkable to those of us who have tasted the very real joys of individual self-fulfillment. This is the terrible fear that lurks behind the tremendous response we have to Orwell and Kafka (among others). I have no easy answer. I would like to think that the palaeo-community was not totalitarian even though it was total. But our rejuvenated societies would perhaps be more like their neolithic counterparts, and there are plenty of examples of these that are tyrannous and destructive of the

for utopias. It is part of the human dilemma that communities can become rotten, repressive and repulsive on any scale. Again I can only argue that if the scale is indeed small, the damage is minimized. This point is absolutely basic. We should not be looking for the perfect community, but for one that is *manageable on a human scale*. Evil will exist on this scale as on any other. The only hope is that if the evil itself is small scale, it will indeed be more manageable, in my sense, than the large scale and totally destructive evils we are now faced with. To be "more human" is not to approach close to some ethical idea of goodness in the usual sense. It is about scale versus survival, not about goodness versus evil. The argument then that palaeo-society can have evil anti-individual aspects is accepted. It is simply declared irrelevant to the present argument which is not about social perfection. At its best, palaeo-society achieves the communalistic ideals while enhancing individual lives; at its worst it perverts both. In either case it does not threaten the survival of the species or make social life an inhuman nightmare.

One of the most attractive of modern utopias, Ursula LeGuin's *Always Coming Home*, does attempt to portray a small-scale social system deliberately based on the neolithic model provided by the Pueblo Indians her father (A. L. Kroeber) knew so well. To their basic system of small, informally linked villages, with matrilineal clans and guild-like ritual and occupational groups, she adds electricity, a dash of geomancy, horse-drawn railways, and a kind of controlled literacy. While she does not make the error of having a society of total goodness, neither does she admit the possibility of true badness: there is no witchcraft, for example, without which the actual Pueblo societies would be unrecognizable. Hers is, after all, a frankly utopian picture, and if one wishes to see the best we could hope for, this is about as good as it could get.

We Have Met the Enemy and He Is Us

On balance? Taking a hard look at the situation I am more pessimistic than optimistic. The brain is in some ways its own worst enemy. Its capacity for illusion and self-delusion, while an evolutionary advantage to "primitive" hunters (as Carveth Read saw in 1920), turns into a terrifying suicidal capacity in (post) industrial society. No wonder people turn again to cults, to astrology, to

in 1920), turns into a terrifying suicidal capacity in (post) industrial society. No wonder people turn again to cults, to astrology, to magic, to hedonistic forgetfulness, to fundamentalism, and to socialism. Socialism is, in its way, yet another cry for a return to the communal ethic. But it fails because like our other modern social philosophies it operates totally within the confines of history and even of industrial history. It has not - except in the insignificant agrarian and anarchistic versions - anything better to offer than more and more industrial progress, with a more equitable sharing of the products of the rape of the earth. It is a prisoner of the assumptions of progress and a leading example of the power of technological hubris. It holds out millenarian hope, and people will cling to this as they cling to the possibility of intervention by benevolent aliens. Both are about as likely to succeed in saving us from ourselves.

Since the beginnings of civilization we have known that something was wrong: since the Book of the Dead, since the Mahabharata, since Sophocles and Aeschylus, since the Book of Ecclesiastes. It has been variously diagnosed: the lust for knowledge of the Judaic first parents; the hubris of the Greeks; the Christian sin of pride; the Confucian disharmony with nature; the Hindu/Buddhist overvaluation of existence. Various remedies have been proposed: the Judaic obedience; the Greek stoicism; the Christian brotherhood of Man in Christ; the Confucian cultivation of harmony; the Buddhist recognition of the oneness of existence, and eventual freedom from its determinacy. None of them have worked. (Or as the cynic would have it, none of them have been tried.) The nineteenth century advanced the doctrine of inevitable progress allied to its eighteenth-century legacy of faith in reason and human perfectibility through education. We thought, for a brief period ("recent history"!), that we could do anything. We can't. But it comes hard to our egos to accept limitations after centuries of "progress." Will we learn to read those centuries as mere blips on the evolutionary trajectory? As aberrantly wild swings of the pendulum? As going too far? Will we come to understand that consciousness can only exist out of context for so long before it rebels against its unnatural exile? We might, given some terrible shock to the body social of the species, as Marx envisioned in his way. (Thus returning us to our state of *Gattungswesen* - species-

also never recover sufficiently from the shock to form the classless, non-industrial communities that were the - albeit vague - Marxian dream of the communalistic future; a dream which is as embarrassing to his followers as Christ's egalitarian pacifist dream has been to the Christian nations.

Being on the side of Man, unfortunately, requires more than just good will. And if Man won't be on his own side, that's his privilege as an intelligent, rational, self-conscious, culture-bearing creature, who has passed beyond the grubby necessities of natural selection to bigger and better things. For as so many well-meaning commentators have so proudly and earnestly proclaimed, He is unique.

O nimium caelo et pelago confise sereno,
nudus in ignota, Palinure, iacebis harena.
    Virgil, *Aeneid*, V.

REFERENCES

(Where the reference to a writer's work is generic, no citation is included; only works specifically referred to in the text are listed here.)

ADAMS, HENRY
1913 [1904]   *Mont Saint Michel and Chartres.* Boston: Hougton Mifflin.

BELL, DANIEL
1973   *The Coming of Post-Industrial Society.* New York: Basic Books.

BOCK, KENNETH
1980   *Human Nature and History: A Response to Sociobiology.* New York: Columbia University Press.

BRADLEY, A.H.
1876   *Ethical Studies.* Oxford: Oxford University Press.

EHRENFELD, DAVID
1978   *The Arrogance of Humanism.* New York: Oxford University Press.

FOX, ROBIN
1983   *The Red Lamp of Incest: An Inquiry into the Origins of Mind and Society.* Notre Dame: University of Notre Dame Press.

FRIEDMAN, JOHN
1979   *The Good Society.* Cambridge Mass.: The MIT Press.

GEERTZ, CLIFFORD
1973   *The Interpretation of Cultures.* New York: Basic Books.

GOODMAN, PAUL
1960   *Growing Up Absurd.* New York: Random House.

HAVELOCK, ERIC
1963   *Preface to Plato.* Cambridge Mass.: Harvard University Press.

HOWARD, MICHAEL
1961   *The Franco-Prussian War*. London: Rupert Hart-Davis.

KROPOTKIN, PRINCE PETER
1903   *Mutual Aid: A Factor of Evolution*. New York: McClure Phillips & Co.

LeGUIN, URSALA K.
1985   *Always Coming Home*. New York: Harper and Row.

LESSING, DORIS
1987   *The Sirian Experiments*. New York: Knopf.

LÉVI-STRAUSS, CLAUDE
1967   *The Scope of Anthropology*. London: Cape Editions.

MORRIS, DESMOND
1969   *The Human Zoo*. London: Cape.

ORWELL, GEORGE
1949   *Nineteen-Eightyfour*. New York: Harcourt Brace.

PFEIFFER, JOHN
1982   *The Creative Explosion*. New York: Harper and Row.

POLANYI, KARL
1957 [1944]   *The Great Transformation*. Boston: Beacon Press.

READ, CARVETH
1920   *The Origin of Man and His Superstitutions*. Cambridge: Cambridge University Press.

REDFIELD, ROBERT
1953   *The Primitive World and its Transformations*. Ithaca: Cornell University Press.

SOROKIN, PITIRIM
1937-41   *Social and Cultural Dynamics*. (4 vols.). New York: American Book Co.

TIGER, LIONEL & FOX, ROBIN
1972   *The Imperial Animal.* New York: Holt Rinehart and Winston.

TÖNNIES, FERDINAND
1887   *Gemeinschaft und Gesellschaft.* Leipzig: Fues Verlag (R. Reisland).

TURNBULL, COLIN
1965   *Wayward Servants.* London: Eyre and Spottiswoode.

TURNER, VICTOR
1983   "Body Brain and Culture" in *Zygon: Journal of Religion and Science.* 18: 221-45.

CONSCIOUSNESS, RELATIVISM AND UTOPIA

Richard Handler
Department of Anthropology
University of Virginia

Cultural anthropologists committed to the notion that humanity's uniqueness is to be found in the use of symbols have often been particularly resistant to the consequences of the evolutionary perspective that Robin Fox presents in "Consciousness out of Context." Nonetheless, Fox's work challenges symbolic anthropologists, and social philosophers in general, to rethink their first premises in the light of the arguments and extraordinarily broad range of data about human evolution that he brings to bear on the issues at hand. The following comments represent my attempt to grapple with that challenge, organized in terms of three topics: (1) cultural relativism and the virtues of the palaeoterrific, (2) the nature of the human mind, and (3) utopianism and rational social planning.

Cultural Relativism and the Virtues of the Palaeoterrific

Fox celebrates the virtues of the "palaeoterrific" society. Along the way, he castigates social theorists, *even those who proclaim themselves to be relativists*, for clinging to what he calls an "eighteenth-nineteenth century social evolutionary way" of evaluating the merits of other cultures in comparison to the culture of the modern West. "I ... object," he tells us, "to the notion of 'progress'," and he adds that "It is nowhere written that it is better ... to be a Western-Rational-Scientific-Literate person than to be a palaeolithic Hunter-Shaman-Warrior-Artist-Poet." He even suggests that modern social theory is nothing other than our version of tribal mythology: both, that is, are imaginative ways to make sense of the cosmos and humanity's place within it.

I believe I will be in agreement with Fox if I elaborate his remarks in the following sense: Most cultural relativism is a cheap, surface relativism in which we say, easily and without thinking the proposition through, that each culture is as good as every other.

Such an attitude accords well with the modern individualism that permeates all our thinking. Cultures, in this formulation, are like individuals, all free and equal. Such an attitude also accords with an equally deep-seated belief in our own superiority and in the benefits of modern rationality and progress. After all, anthropology is thought to be the science that enables *us* to understand *them*. We merely delude ourselves, I think, when we pretend that anthropology allows us to understand other peoples on their own terms. The best anthropology approaches such a goal, but the anthropological project is, first and foremost, a Western, scientific project which almost in spite of itself demands that we conceptualize others in our terms. In other words, anthropological relativism is not incompatible with an unthinking belief in the superiority of our own ways of thought.

Looking beyond anthropology, to less arcane discourses, I think it is obvious that most of us believe that those non-Western, primitive others are deprived, impoverished, inferior in relation to us. (Think, for example, of the stubborn resistance of some of those clamoring for increased cultural literacy, defined in part in terms of acquaintance with artistic and literary masterpieces, even to consider the possibility that so-called primitive art and traditions might rank with the Dantes and Beethovens of the world.) Thus I am sympathetic to Fox's playful description of the intellectual satisfactions that would have been available to him had he been a palaeolithic hunter. I do not think he overstates the case at all in castigating what he calls "our arrogance."

Having agreed thus far with Fox, I must point out two ways in which my relativism differs from his celebration of the palaeoterrific. First, Fox not only celebrates the palaeoterrific, he argues that it represents "the basic pattern" of human society. He in effect reverses our ethnocentric and arrogant ranking of societies to say that it is *we* who are degenerate, *they* natural and healthy. By contrast, I would like to avoid such ranking, and in what follows will challenge Fox's notion of the basic pattern.

Second, Fox rightly (in my opinion) castigates our conviction in the superiority of our rationality; he satirizes, for example, the pretensions of social science and social theory. But, I would ask, *on what grounds does he exempt evolutionary biology from his attack on modern modes of thought?* To this issue, also, I shall return below.

The Nature of the Human Mind

Fox gives us a glimpse of what he means by "the basic pattern" of human society when he mentions "organic extended kinship groups." In his book, *The Red Lamp of Incest*, he has described in detail the social properties of such groups: hunting and gathering, a sexual division of labor with rules for the sharing of the products of labor, rules to regulate sexuality, hierarchies of dominance (among both men and women) which have consequences for each individual's ability to reproduce, and so on. In *The Red Lamp*, as in the paper under discussion, Fox argues that the human species *is adapted* by millions of years of evolution to live in this kind of palaeoterrific society. In his paper, he describes our "primate brain re-tooled by predation," a brain that is "geared for a particular range of adaptive responses." He writes also of the "pristine behaviors surrounding its primary functions: survival and reproduction" - and of the "balance between the organism, the social system, and the environment" that existed in the time of the palaeoterrific.

Behind these descriptions of the basic human society there is a careful discussion of the nature of the human brain, that is, of the brain as an organ adapted by evolution to function and respond in certain ways (precisely those ways that result in palaeoterrific society). In this discussion, as developed in *The Red Lamp*, Fox refuses to separate the conceptual abilities of the brain from its emotional propensities. As he puts it on p. 173:

> We ... have at least two fundamental processes going on here [in the brain] .... There is the urge to classify - the intellectual process - and the urge to interdict - the emotional.

And on the next page:

> ... we easily learn fear, aggression, love, language, incest avoidance, attachment, and altruism. We also learn to categorize, interdict, exchange, and make rules - to employ the whole range of mental activities, whether we dub them intellectual or emotional, that we can see are the outcome of our evolution....

In developing such arguments, Fox is arguing against those philosophies of Man, so central to Western thought, that separate human beings from all other animals on the basis of our intellectual (or, in Christian terms, our spiritual) powers. Some symbolic anthropologists, philosophers, linguists, and psychologists have been so impressed with the apparent primacy of symbols and symbolling in human life that we have defined humanity in terms of them - that is, in terms of the conceptual, classifying propensities of mind. Fox amply discusses the importance of symbolling and classification in human life and thought, but he refuses to give them precedence over other mental functions, refuses, that is, to divorce intellect and emotion. Human beings by the nature of their minds classify, he agrees, but one of the things that they *universally* classify, he argues, is kin (*The Red Lamp*, 183); and, he continues, not only do they *classify* kin, but, motivated by deeply-rooted emotional attitudes, they regulate sexual access to some of the people thus classified. In other words, the structure of palaeoterrific society - the sharing, the sexual division of labor, the control of sexuality and the dominance hierarchies, all so primate-like but so unlike the other primates - is an outgrowth of *the nature* of the human mind, the mind that classifies emotionally.

To this argument I would respond that because human thought and experience are dependent upon and constructed in terms of symbols, concepts, and categories, human beings *as thinkers* are no longer bound by their evolutionary history. I agree with Fox in deeming our conceptual abilities to be the product of the evolution of the brain; I agree, therefore, that our human potential is built into our nature, as it were. But given those conceptual abilities, given an animal that responds not directly to the world out there, whatever that is, but to a world experienced in terms of symbolic thought - then our responses to the world are no longer limited by our evolutionary nature. And this argument is intended to comprehend our thinking about the experiencing of "the emotions," whatever they in their raw state might be. As Clifford Geertz has written,

> ... there is no such thing as a human nature independent of culture. Men without culture ... would be unworkable monstrosities with very few useful instincts, fewer recognizable sentiments, and no intellect ...

> Our ideas, our values, our acts, even our emotions, are, like our nervous system itself, cultural products....
> *The Interpretation of Cultures* (pp. 49-50)

If, then, even our emotions exist for us only as our culture - or, more precisely, as the culturally patterned conceptual functioning of our minds -constructs them, then I no longer know what sense to attribute to Fox's notion that such universal emotions as aggression, altruism and love, evolutionarily programmed into our brains, constrain the particular pattern of social interactions appropriate for humans. It may well be that our wandering beyond the confines of palaeoterrific society is tragic, but given our thinking minds, such wandering would seem to be fated and unavoidable. In other words, our human nature does not uniquely fit us to live in a particular kind of society, be it the palaeoterrific or some other, but forces and enables us to invent and live many social experiments, each of which is called "a culture" by relativists.

Fox seems to agree to some such vision of humanity's fate when he asks whether we could have stopped there, in the palaeoterrific, and answers "I very much doubt it." The cultural imagination, he tells us, is the product of our brains and, as such, "correspond[s] to *something* in the creature." But if that is so, I don't see how Fox can separate out that part of human nature which adapts us, in his view, to the palaeoterrific from that other aspect of our nature which allows us to imagine and create cultures beyond the palaeoterrific. Both the palaeoterrific and the cultures beyond or beside it are made possible by the human nature of our brains, and all, I would argue, are constructed by a cultural imagination, not by the force of instincts and emotions that continue to speak to us through millions of years of evolutionary history.

With regard to the question of what took us beyond the palaeoterrific, I do not accept Fox's "social density" argument. He tells us that "population was squeezed into the Middle East and southwestern Europe by the ice, and the unprecedented social density thus created led to a burst of self-conscious activity evidenced by the fantastic art of the period." In contrast, I would argue that self-consciousness and, more particularly, the ability to reflect upon our situation, our culture, our mode of thought, means that creative re-interpretation of any and all cultural situations is inescapable for human beings, the thinking, symbolling animals. We cannot stand still, in the palaeoterrific or anywhere else, because we are condemned perpetually to conceptualize *and to reconceptualize* our condition. Thus it is not enough to suggest that the self-conscious creation of post-palaeoterrific culture was the product of, almost a mechanical response to, increased population density. Let us say, rather, that self-consciousness is in "the nature" of *Homo sapiens sapiens*.

Utopianism and Rational Social Planning

Fox insists he is not a utopian, and I am inclined to agree with him. After all, he points out that life in the palaeoterrific included war, aggression, illness, the precariousness of the hunt. He explicitly distances himself from noble-savage rhetoric. And he asserts that "our palaeo-nature is a mixture of good and evil," and that, therefore, we cannot expect human societies, constructed upon the basis of that palaeo-human nature, to be without expressions of the bad.

However, though he is not a utopian, Fox admits to kinship with the "'small is beautiful' crowd." I wonder if we can fairly place him in a pastoral tradition, or with those thinkers in the Western tradition who have looked back to a golden age? What comes most immediately to mind, however, is Rousseau's vision of the first human societies, as described in his *Discourse on the Origin of Inequality*. It will be recalled that Rousseau criticized social-contract theorists, particularly Hobbes and Locke, for projecting 17th-century English culture back into the state of nature. Natural man, Rousseau argued, had neither the means nor the desire to construct a social contract, because natural man could only have been an animal without thought and language. And without these, natural man would not lust after and create property, would not communicate with other humans, and certainly would not fabricate a social contract. For Rousseau, the origins of society depended on the acquisition of human thought, defined explicitly in terms of symbolic classification.

Trying, without the benefit of evolutionary theory, to imagine the pre-history of humanity, Rousseau considered the difficulties that would have confronted a non-symbolling animal in the process of acquiring symbolic ability:

I am so aghast at the increasing difficulties which present themselves, and so well convinced of the almost demonstrable impossibility that languages should owe their original institution to merely human means, that I leave, to anyone who will undertake it, the discussion of the difficult problem, which was most necessary, the existence of society to the invention of language, or the invention of language to the establishment of society. (p. 63)

Thus did Rousseau, not wishing to appear to argue against the divine creation of human language, abandon his speculations about its origins. He had, after all, established to his satisfaction the foundational place of language in human society, and that was enough. For him it was sufficient to argue that once humans had acquired language, they would begin to live together in little societies. And these first societies were, according to Rousseau, "the very best man could experience."

Like Fox, Rousseau saw the development of civilization beyond early society to be a history of retrogression rather than progress: all "advances" beyond it, he wrote, "have been apparently so many steps towards the perfection of the individual, but in reality towards the decrepitude of the species" (p. 83). *Unlike* Fox, however, Rousseau attributed the decline of humanity not to some mechanical process such as social density, but to the very thing which, in Rousseau's scheme, made humans human: the acquisition of language. Without language, humans could not classify experience, and without the ability to classify discretely experienced pieces of the world as "same" and "different," the human animal could not compare things or people and differentially value them. *With* language, however, comes the inevitability of invidious comparisons: a man whose instinctive sexual needs had once been satisfied by any woman will now, with language, know that this woman and that woman are both women, hence the same, but also different in their possession of those differentially valued attributes that add up to "beauty." Rousseau described the general situation this way:

Whoever sang or danced best, whoever was the handsomest, the strongest, the most dexterous, or the most eloquent, came to be of most consideration; and this was the first step towards inequality, and at the same time towards vice. (p. 81)

So, according to Rousseau, it is precisely that which makes us human - the possession of language and symbolic ability - that expels us from paradise and makes human history tragic rather than progressive. There thus can be no "steady state" of human society, such as Fox's paleoterrific, for, as I have argued earlier, our human nature, defined as it is in terms of symbolic ability, condemns us perpetually to rethink our situation.

In rethinking our situation, can we plan, deliberately, to return to the small-scale societies that both Rousseau and Fox celebrate? At the end of his paper, Fox raises the question, and ventures a somewhat ambiguous answer, what he calls "a realistic, piecemeal social engineering" that will yield, not utopia, but social arrangements more attuned than our present state to our palaeo-human nature. The problem with this solution is that it depends upon a consistently exercised rationality that would seem, at least in Fox's vision, to be beyond human abilities. After listening to him castigate our modern faith in progress and rationality - and be it recalled that I agreed with Fox's critique of Western arrogance - one wonders how he can turn to the rationality of social engineering to rescue us. Or, as I put it earlier, given his refreshing dismissal of most social philosophy, he ought at least to argue for the privileged status that he accords to evolutionary biology, the sole discipline, apparently, capable of rescuing humanity from itself.

REFERENCES

FOX, ROBIN
1983   *The Red Lamp of Incest: An Inquiry into the Origins of Mind and Society.* Notre Dame: University of Notre Dame Press.

1988   "Consciousness Ot of Context: Evolution, History, Progress and the Post-Post-Industrial Society." This volume.

GEERTZ, CLIFFORD
1973   *The Interpretation of Cultures.* New York: Basic Books.

ROUSSEAU, JEAN-JACQUES
1973   *The Social Contract and Discourses.* Trans. G.D.H. Cole. London: Dent.

SCIENCE AND HUMAN NATURE

George Klosko
Department of Government and Foreign Affairs
University of Virginia

Reading professor Fox's paper, I was strongly reminded of Rousseau. In the *Second Discourse*, Rousseau, of course, attempts to distinguish what is natural from what is artificial in man's nature. Rousseau argues that man, as he comes from the hand of nature, is, to put it simply, an animal. He lives a solitary existence in the woods, but because his mind is undeveloped and his needs are few, he is content. Rousseau writes:

> I see him satisfying his hunger at the first oak, and slaking his thirst at the first brook; finding his bed at the foot of the tree which afforded him a repast; and with that, all his wants are supplied. (1973:47)

According to Rousseau, the three basic constituents of natural man's psyche are the desire for self-preservation (*amour de soi*), compassion, and perfectibility, the ability to develop and change over time. One generation of cats is much like the one that preceded it, but this is not always the case with human beings.

The *Second Discourse* relates the story of how man lost his original innocence and came to be the corrupt and sorry creature that we see around us. For Rousseau, this change was comprised of a series of distinct steps. As man's circumstances changed and his existence was imperiled, he was forced to cooperate with other men in order to satisfy his needs. As a result, his ability to reason developed. But the appetites of a reasoning creature are unlimited. Man's needs expanded; most important, he acquired the need to be thought well of by his fellows, vanity (*amour propre*), which eventually came to overshadow all his other needs, and man's social condition took on hellish aspects. Rousseau's proposed solution is presented in *The Social Contract*. He advocates a return to simple, agrarian societies, in which various forms of social inequality will be minimized. People will become virtuous, focusing their psychic energy on the good of the community, rather than their own particular interests. Rousseau does not believe that a complete return to man's natural condition is possible. The ideal society sketched in *The Social Contract* represents the greatest *possible* return - and even this would require a miracle to accomplish.

Professor Fox, too, attempts to distinguish what is natural and what is socially induced in human nature. Because he believes that man is in a sorry state in existing society, for him too the account of man's nature has strong normative implications.

The basis for Professor Fox's analysis is a study of human origins, "the more than five million years of natural selection" that have made man what he is (Fox: this volume). According to Professor Fox: "anything but an evolutionary view of modern man"(3) will not be sufficient to diagnose man's present plight. Evolutionary biology reveals the existence of a primordial stage, from which human beings have evolved, "and in which we are *supposed* to exist" (8; his emphasis). The small organic groups in which men lived at this stage - comprised of roughly forty persons - catered to "the whole range of human needs and satisfactions for each and every member"(8). The upper-paleolithic era presented a balance between man's organism, his social system, and his environment (13). Apparently in all seriousness, Professor Fox refers to it as the "paleoterrific" era (12).

For Professor Fox, then, like Rousseau, subsequent history has not meant progress. Modern society is not natural; it is founded upon a violation of human nature. Like Rousseau, Professor Fox traces the "slow succession" of steps that have removed man from his natural condition (Rousseau 1962: 103). Our discontent in society reveals a deep seated longing for a simpler time. As Freud says, when we consider how unsuccessful man has been in regulating social interactions and the great pains that have resulted, "a suspicion dawns on us that ... a piece of unconquerable nature may lie behind ... a piece of our own psychical constitution" (Freud 1961: 33).

Professor Fox has no easy solution to the difficulties that he recounts. Again the comparison with Rousseau is instructive. His social theory recommends a return to man's natural state, to "palaeo-morality," the ideal that it embodied. Though human beings have changed as they have evolved and a complete return is no longer possible:

> ... we must try to restore such of it [palaeo-morality] as we can in relatively limited circumstances, recognizing that we have long ago passed the point where we could insist on absolute and universal application (28).

> The point is absolutely basic. We should not be looking for the perfect community, but for one that is *manageable on a human scale* (29; his emphasis).

His specific recommendations are to some extent familiar, as having been advocated by the "small is beautiful" crowd:
> We *can* attempt a realistic, piecemeal social engineering of our over-individualized, over-rationalized, over-bureaucratized and altogether too large societies, to produce small-scale institutional settings where the paleao-morality would have more chance to flourish (28; his emphasis).

Professor Fox and Rousseau are also similar in believing that their specific claims about human nature are rooted in science. In Rousseau's *Second Discourse* the scientific material is presented mainly in the notes (too often condensed or omitted from translations of the work), in the form of anthropological and zoological evidence that he has assiduously (if somewhat credulously) collected from an impressive variety of sources. Professor Fox's claims to scientific status are rooted in evolutionary biology. He states that his findings rest upon "a hard-science view of the evolution of human behavior"(14). And he criticizes the "small is beautiful crowd" for having less substantial ideals: "they have no basis in evolutionary biology to direct them positively"(26).

Though I find much in Professor Fox's presentation persuasive and attractive, I believe that there are difficulties with his account of its scienticity. To the extent that Professor Fox's paper presents more than pleasant myth, this is because his account of the paleolithic era enjoys scientific support. But we should be quite clear about how much of his argument can be termed "scientific." Though aspects of Professors Fox's account of the paleolithic era are undoubtedly based on scientific evidence, he encounters severe difficulties in moving from these to his positive recommendations concerning modern society.

To begin with, I would question Professor Fox's claims for the "naturalness" of life in paleolithic societies. Professor Fox of course holds that such an existence was paradisiacal in comparison to the complex, difficult conditions of modern societies. But "natural" is a normative rather than a scientific concept; translated, the term could just as well mean "good" or "desirable," which are of course more overtly normative concepts. Professor Fox has little evidence that paleolithic life was as happy as he depicts it. His account here is little more than pleasant myth; his discussion of his own well-adjusted life in paleolithic times is clearly a fantasy. Now, one way to shore up his position would be to argue that people are happier in smaller, less complex societies than in the modern world. Rousseau argues along these lines in the *Second Discourse*. Rousseau says that one never finds individuals from primitive, tribal societies, which he believes to be the happiest of which men are capable, deserting them for more advanced societies, while there are numerous cases of members of more advanced societies leaving them for the less advanced. Rousseau discusses such a case in the *Second Discourse*, that of a South African youth who had been removed from his tribal society and raised with the accouterments of European civilization, who cast the trappings of civilization aside to return to the bush (1962: I, 219). Arguments based on comparative anthropology are doubtless common and perhaps persuasive. However, if Professor Fox wishes to argue in this way, his argument would be made on the basis of sociological and anthropological comparison. Little would depend upon the findings of evolutionary biology.

I believe that there is an implicit argument in Professor Fox's paper, comprised of four main steps. He argues from (1) the fact that human beings evidenced certain characteristics during the paleolithic era, to (2) these are basic to human nature, and (3) these are therefore desirable, so (4) they should be fostered in modern societies. This is an interesting argument but it must be developed a great deal more before it could bear up under scrutiny. For our present purposes it should be noted that this should *not* be described as a scientific argument. It depends on crucial value judgments in the extrapolation of certain features of the paleolithic era as basic in step (2) and the postulation of their desirability in step (3). Because of the normative nature of these steps, the scientific element is largely muted in Professor Fox's paper.

In general, accounts of human nature that claim to be rooted in science are inherently suspect. Many of the great figures within the tradition of political theory have argued from specific accounts of human nature to specific moral and political conclusions. In the cases that are of greatest interest here, one finds an argument that consists of two major components. We have, first, an account of man's nature, frequently in the form of a depiction of man's natural state, the "state of nature." This account of the state of nature then develops into a description of the difficulties encountered in that condition which make political arrangements necessary. Because of the "inconveniences" of the state of nature (Locke: §127), people were forced to band together and erect governments, thereby giving birth to civil society. The specific forms that these political arrangements took are justified by the need to remedy the particular problems of the state of nature.

An argument along these lines is clear in Rousseau, whose account of man's natural condition in the *Second Discourse* provides moral justification for the specific political forms recommended in *The Social Contract*. It is perhaps more clear in Hobbes and Locke, both of whom argue that, if one wishes to see the particular government that we need, one should postulate a condition in which men live without government, i.e., a state of nature. Their accounts of the specific difficulties in their respective states of nature are used by Hobbes to justify effectively unlimited government, and by Locke to support limited government. It is not clear that Hobbes and Locke took their states of nature seriously as historical accounts of man's actual condition. Though Hobbes makes strong claims to pursuing a rigorous scientific method, and though both he and Locke present some evidence for the validity of their accounts, they appear to be content with states of nature that are hypothetical, analytical constructs, as well they should. In reading Hobbes and Locke, one sees clearly that their states of nature are not actually the premises from which their arguments emerge. Rather, both thinkers have specific political agendas and so wish to uphold particular conceptions of legitimate political authority. Their political conclusions should be viewed as primary, with their accounts of the state of nature (and so of human nature) constructed after the fact, deliberately shaped to justify the desired political outcomes.

In the case of Locke (again, perhaps unlike Hobbes) it is clear that we do not have a neutral, scientific account of human nature. Because Locke does not make strong claims concerning the scientific status of his arguments, this is probably not a ground for criticism. The case is similar with Rousseau. Though he has some scientific pretensions, they are rather weak, and again he escapes criticism. But because Professor Fox makes stronger claims, he should be criticized.

I believe that there is no neutral, scientific account of human nature. To some extent my claim rests upon semantic points, definitions of "neutral" and "science." These I will avoid. But granted the fact-value distinction, postulated in some reasonable manner, a conception of human nature can be seen to fall on the "value" side - regardless of the extent to which it is supported by scientific evidence and scientific arguments.

Of necessity, any account of human nature will make distinctions in regard to man' needs and capacities. Some will be designated as "natural" and others "unnatural" or "artificial." These distinctions have strong normative implications, as needs of the former sort are granted a *prima facie* claim to satisfaction; those of the latter sort receive a presumption against satisfaction (see S. Lukes 1974: 185-86, and C. Taylor 1973).

Similarly, an account of human nature will be committed to a view of human potential. A theory will not only describe man as we see him in existing society, but claims will be made about what man is capable of in other, reformed societies. These claims have significant political implications: one's range of political alternatives is determined by what one views as possible. It has been clear since Mannheim's *Ideology and Utopia* that political partisans (consciously or unconsciously) will often attempt to rule out various possible courses of action by describing them as utopian, as beyond the range of human possibility.

The connections under discussion are clearly seen in Marx. Marx's strong (largely implicit) claims about the possibility of human perfection in the future, communist societies rest upon his postulate of virtually unlimited human potential, and this because man's nature is largely unfixed. According to Marx, man takes on the characteristics of the society in which he is raised. As he writes in the sixth Thesis on Feuerbach: "[T]he human essence is not some abstraction inherent in each single individual. In its reality it is the ensemble of social relations" (Marx 1978: 145). That human nature will *change* under communism is made quite clear in Lenin's exposition of the Marxian theory of revolution in *State and Revolution*.

Then again, to take another example, according to the conservative Freudian view, human nature is fixed, with a large store or primordial aggressiveness. Radical reform is therefore beyond the realm of possibility. The task of society is not to reform man, but to control his destructive tendencies. Similar views of human nature and so similar political precepts have been defended by other conservative thinkers, from St. Augustine to Hobbes.

It is worth noting here that Marx and Freud, with all their differences, both claimed to be expounders of science.

Perhaps a proponent of a scientific view of human nature would respond to the points that I have been making along the following lines. Professor Fox, for instance, could say that his view is scientific (at least to a large extent) because his view, unlike those of the thinkers I have noted, is rooted in incontrovertible facts, the findings of evolutionary biology, concerning how man lived and evolved.

I have grave doubts about the force of such an argument. When we speak of a "fact," we have in mind something that is firmly grounded, generally in experience, and so something in the existence of which we can have confidence. I believe that the kind of evidence that Professor Fox would be able to produce to support his claims would fall far short of "facts," in this sense. The main problem with facts bearing upon views of human nature is that they are inevitably selective. A view of human nature cannot possibly be based upon a complete enumeration of all the evidence, all the "facts," bearing upon its validity. There is an infinite amount of evidence relevant to any view. Any particular view of human nature will concentrate on certain aspects of human experience, which are at least implicitly deemed important, and pay less attention to others, which are unimportant. Any particular view will pursue particular questions - and not others - from a limited number of possible approaches. The resulting view of human nature is less a complete - or even a neutral - recounting of facts than a snapshot. What we have is a particular selection of the evidence, worked up to produce a coherent, (generally) intuitively plausible picture. The problem with such views is that they resist easy disconfirmation. While part of what we mean by a "fact" is that it is subject to disconfirmation by conflicting "facts," overall views of human nature are generally sufficiently flexible to accommodate any observations that can be brought against them.

When Professor Fox looks back and sees happy primitive man in his natural environment, living according to palaeo-morality, he is, of course, presenting a selective view. It seems to me that the challenging questions bearing upon his theory of human nature would concern justifying his approach and his evidence. Why consider the evidence of paleolithic society and not other sources? Why choose the particular aspects of paleolithic society that he discusses? These questions Professor Fox does not address. Looking at similar paleolithic societies, Marx would see poverty and exploitation, and so the need to develop the means of production that slumber in nature's womb. Freud would see the eternal battle between eros and thanatos, between civilization and the aggressive instinct's desire to break free.

Which one of these views is the correct one? I believe that it is not possible to say with certainty. What must be emphasized here is that the question of correctness is not a scientific question. The problem is not to devise empirical tests and then to assess the evidence against them. Rather, it is a question of values. The Marxian, the Freudian, and Professor Fox assess the evidence from different points of view. They not only draw value conclusions from their views of the evidence; but their values influence the ways they approach the evidence. In choosing to look at certain aspects of paleolithic societies rather than others, in choosing to employ certain means of observation rather than others, these thinkers approach their "scientific" observations in such a way that certain conclusions are pre-ordained. Again, as a rule, the conflicting views of human nature that their observations would support would be sufficiently flexible to absorb virtually any conflicting evidence.

According to J. S. Mill, the great political theorists are "one-eyed men." Each has left us with a particular view of the nature of man and of society. Each emphasizes certain aspects and downplays others. Their theories leave us with conflicting portraits of human experience that are often impressive, and sometimes profound. But such theories should not be mistaken for science.

REFERENCES

FOX, ROBIN
1988 "Consciousness out of Context: Evolution, History, Progress and the Post-Post-Industrial Society." This volume.

FREUD, SIGMUND
1961 *Civilization and Its Discontents.* J. Strachey, trans. New York: W.W. Norton.

LOCKE, JOHN
1985 *The Second Treatise of Government.* New York: Liberal Arts Press.

LUKES, STEPHEN
1974 "Relativism: Cognitive and Moral," *Proceedings of the Aristotelian Society*, Supplementary Volume 48.

MARX, KARL
1978 "Theses on Feuerbach" in *The Marx-Engels Reader*, 2nd ed. New York: W.W. Norton.

ROUSSEAU, JEAN-JACQUES
1962 [1915] "Second Discourse" in *The Political Writings of Jean-Jacques Rousseau.* C.E. Vaughan ed. and trans., 2 vols. New York: Cambridge University Press.

1973 *Discourse on the Origin of Inequality* in *The Social Contract and Discourses.* G.D.H. Cole, J.H. Brumfitt, and J.C. Hall, eds. and trans. London: Everyman.

TAYLOR, CHARLES
1973 "Neutrality in Political Science" in *The Philosophy of Social Explanation.* A. Ryan, ed. Oxford: Oxford University Press.

A COMMENTARY ON FOX'S DIAGNOSIS OF THE HUMAN CONDITION

Julian N. Hartt
Kenan Professor of Religious Studies, Emeritus
University of Virginia

Backward, turn backward, O Time in
   Your flight,
Make me a child again just for tonight!
Backward, flow backward, O Tide of
   The years!
I am so weary of toils and tears
   All in vain--
Take them and give me my childhood again!

           Elizabeth A. Allen
          (Bartlett, 1937, p. 595)

Introduction
Lament for the loss of community is an unmistakable and poignant note in contemporary critiques of modernity and emerging post-modernity. Prescriptions for the recovery of this treasure sometimes accompany such critique. But one must be prepared to look to the past to catch strongly affirmative feelings abut such prescriptions, optimistic sentiments, blessed assurances that it is not too late to realize essential humanness in recovered community. Without those assurances critiques of the present human order are something like wakes in which familial and friendship ties are warmly renewed. No one expects that warmth of sentiment to re-animate the corpse. Were so bizarre a prospect seriously entertained it would almost certainly cast a pall over an ambiguously lugubrious occasion.
    Nevertheless, prescriptions for the recovery of lost community are still forthcoming. The day is all but spent but it is not yet too late. If only we will highly resolve to do thus-and-so ...

Who are these *we*? People who take the diagnosis to heart and thus recognize the moral-spiritual maladies attacking human being in its modern formations and now indeed threaten to stamp *finis* on the entire human enterprise. Who are these *we*? People with enough courage to attempt the all-but-impossible; not to save the world but to create an oasis of sanity and health and human warmth in the desert fashioned by the forces of history since the Fall. The really good life is not an impossible dream, quite. The retrieval of that life would have to be something like a family affair. It were sheerest folly to count on massive social forces, no matter how high-minded the directors of same, to bring in that good life.

Professor Fox is as sure as any of the nay-sayers against modernity that abysmally wrong turns in human history have landed the present age in a quagmire from which there may not be extrication for any significant fraction of humanity. The epoch which produced the genocidal holocaust may top it off with the nuclear apocalypse. Professor Fox does not indicate whether he would see in such a *denouement* a stroke of justice. A pall of inevitability hangs over his view of the future. But the inevitable is different from the inexorable. The latter suggests a decree that cannot be (a)voided, "a judge who cannot be placable" (*Cassell's Latin-English Dictionary*, Macmillan, 1987, p. 85). Professor Fox's persistent wariness about strong value judgments should argue, I suppose, that he would resist the *inexorable*.

In what follows in this commentary I first consider some conceptual puzzles in his provocative essay. This is not a very large concern. It is just the sort of thing philosophers do while waiting for bigger game to show up. Not too big for their weapons and ammunition. It is disquieting to be sitting there with only a squirt gun when a rhino shows up. Rhinos aren't all that playful. Or so it is said.

Bigger game certainly shows up. So in the second part of this commentary I look at Professor Fox's interpretation of history, that is, at his diagnosis of the present age and his critique of soteriologies old and new.

Finally there are his suggestions for mending our ways as the post-modern world heaves into sight and sound - and smell. There may not be a way out. Even so there might be a way of "redeeming the time," if I may be pardoned a little lapse into Christian scripture (*Ephesians* 5:16).

Some Conceptual Puzzles

There are three sets of conceptual puzzles in Professor Fox's essay which I consider here. At the risk of incurring his displeasure I lay these out as dichotomous pairs.

1. BRAIN/MIND
2. TIME/HISTORY
3. VALUE/REALITY

I do not intend that this layout should be viewed as the nose of a metaphysical camel angling for the least protected part of a scientific tent. Perhaps people who ignore metaphysics are bound (fated) to commit it. This is not the occasion for assaying the truth of that proposition; or even for determining whether it is a fugitive scrap of - metaphysics.

1. BRAIN/MIND

Professor Fox in this paper is not much inclined to use *mind*. He makes persistent use of *consciousness*, as in the title of his paper. Perhaps he finds that *mind* is too deeply tainted with intellectualist prepossessions, as Descartes' *cogito* appears to be. But both Descartes and Fox are engaged in thinking, not just in being conscious. Like Fox Descartes thought a lot about the mess western civilization was in (and it was only the 17th century!); and what to do about it.

But what is puzzling about Fox's use of brain/mind? Several things. (a) Does he want us to recognize that consciousness has evolved just as the brain of *Homo sapiens* has evolved from the brain of anthropoids of one kind or another? Modern man does things with his mind that Neanderthal and Cro-Magnon didn't do. Calculus was invented in the 17th century. Cro-Magnon could count, no doubt, whatever the system. What would we gain by saying that calculus *evolved* from Cro-Magnon's system? In respect to almost any human activity one could think that early ("primitive") forms of it are generally a condition of the latter forms, but the weight of that condition varies enormously from one kind of activity to another. To identify a primitive form of an activity as the necessary and sufficient condition for everything in it thereafter begs the question.

So far Professor Fox has not done that. The going gets thicker and more interesting when he imputes to the brain behaviour ordinarily attributed to the mind.

> The brain is in some ways its own worst enemy. Its capacity for illusion and self delusion, while an evolutionary advantage to 'primitive' hunters ... turns into a terrifyingly suicidal capacity in (post) industrial society. (p. 30)

We have no trouble understanding how minds - or selves, persons, if you prefer - can suffer illusion and self-delusion. But the *brain* as a physical organ? Why not put other organs on a comparable basis, and say of the liver that it has a suicidal capacity for over-indulging in alcohol? Or of the penis that it often - perhaps systematically - seeks illusory satisfactions, yearns to be what and where no penis naturally belongs?

Earlier in his essay Professor Fox says,

> We - the creature - produce these fantasy structures - cultures, religions, laws civilizations. We produce them out of the raw material of our speeding brains and ricocheting imaginations.(p. 20)

I do not suppose he means us to infer that these "fantasy structures" are pure fantasies. For then what purchase would science have on reality? Freud exempted scientific rationality from the disease of rationalization. It must be the case, more or less analogously, that Fox believes science is on the right track in its delineation of Upper Palaeolithic Man. We must assume that the essentially human life he discovers in that creation is not simply fashioned from somebody's ricocheting imagination.

Before we leave this first puzzling pair (BRAIN/MIND) I call attention to the sentence in the passage just quoted: "*We* produce them out of the raw materials of our speeding brains and ricocheting imaginations." Compare this with, [the brain] is not an organ of cool rationality: it is a surging field of electro-chemical activity replete with emotion ....(p. 5)

Is the emotion a by-product of that surging electrical field? Or is it a native force, so to speak, responsible for heating-up the brain, making it difficult, but surely not impossible, to play it cool?

I do not want to foul up the operation with metaphysical grit. So I back off the first puzzling pair with an observation - not an argument, metaphysical or otherwise. If a scientific or philosophical appraisal of the human condition seriously intends to come up at the end with prescriptions other than either Epicurean or Stoic despair, then personal agency must be let in on the ground floor. Professor Fox's *we* in the statement quoted above must be taken seriously: "*we* produce" the fantasies; we convert some of the fantasies into plausibilities; we convert some of the plausibilities into truths; truths scientific and philosophical and aesthetic and religious. And, oh yes, historical.

2. TIME/HISTORY

So long as we are content with pre-systematic language the relation of history to time is not very complex. Time is a medium in which all that is finite ("creaturely," in traditional religious language) acts and is acted upon.

System draws on apace when we say that time is the measure of change; so that if there is being that does not change it may be timeless in some interesting way. Of course, as moderns we believe nothing is so interesting as change, probably nothing so real as change; so anybody who wants to put in a good word for changelessness may find it hard to keep the attention of his audience, to say nothing of their sympathy. They will spend a lot of their time and his looking at their watches.

History is human time: passage evaluated as interesting or not, important or not. Accordingly memory is the fundamental human appropriation of passage. And thus the eventual convergence of history-as-lived and history-as-written. We moderns have a deep (perhaps incorrigible) belief that if you can't dig up the documents you don't have *real* history. (In fact have you tried recently to establish personal identity without any document? You may claim existence but you were never born without documentary substantiation.)

Professor Fox has a profound interest in yet another modality of time: cosmic - geological (scientific) time. So system seems to emerge full-orbed. Even so this is a modality akin to commonsense passage. The big difference is the expansiveness of scientific time: it lunges terrifically both towards the infinite and the infinitesimal. It comes in magnitudes intelligible and useful only through very special instruments of observation, the telescope of cosmic range and the electron microscope. To them thanks for light years and nanoseconds.

Professor Fox has a large investment in this time-modality and a very slender one in history. Human time is a very late and short - so far - span as measured against geologic time; as is the age of this planet measured against the age of the cosmic system. Commonsense may rally from the stupefaction induced by the display of these large numbers and ask: "Who is the *measurer*?" Surely not a god - except in his own estimation.

There is another aspect of this all-encompassing time of modern science, and it is of great interest for Professor Fox. Time is not just passage, it is not mindless ceaseless change. Time is an intelligible process. The name of the process is evolution. (When I was a callow youth the philosophical-theological-scientific air resounded with Emergent Evolution. This was a view that systematically linked time with creative process. The backers of EE did not claim that reality was necessarily getting better but it was (is) certainly always getting more interesting.)

As Professor Fox construes the picture human being is in every important respect a product of evolution. A very late product at that. In the next section we shall review his claim that human being was wrench out of the environment for which his brain and its "mechanical appendages" were naturally adapted. Man has done this to himself, apparently under the catastrophic illusion that he could thus improve his position in nature's game, whatever that is. Before proceeding to that let us consider an element in his assessment of history. In what follows Professor Fox is treating "most views of man and society":

... they all suffer from the same problem: they operate from inside "history." They have no view of the place of history in the total story of the human species. (p. 23)

In other words conventional academic social thought fails to take the long view in which "the total story" necessarily resides.

The main question about this claim is, Where does its author stand in order to make it? In terms less tainted with *ad hominem,* how is it possible to stand *outside* of history? Is he prescribing access to truth *sub specie aeternitatis*? Ought we to detect an echo of Socrates in *Theaetetus*: real knowledge of passage presupposes a mind (cognizing subject) that effectively transcends passage?

Far-fetched possibilities, these. A far more plausible one is at hand: (a) infatuation with human time (history) inevitably tails off into philosophic parochialism; (b) hence an illusory sense of the cosmic importance of being human. Adding (a) and (b) we have a demand to reorient the human enterprise in the immense spatio-temporal complexity of nature.

3. VALUE/REALITY

As we are about to see in the third section of this paper Professor Fox has a strong complaint about the present formation of the human condition. Human being isn't what it ought to be; it was much better a long time ago, long in historical, not cosmic, time. Indeed it is possible to read Professor Fox's essay as a sustained value judgment on history for ten thousand years. Hence to the next question: What grounds value distinctions and value judgments? What sort of grip do our values and valuations have on reality.

There are echoes in Fox of Vaihinger's (1852-1933) philosophy of As If. We are bound, that is, to act as if our values, including truth, somehow mesh with reality. At their best our value prepossessions are beneficent fictions.

But I am not sure that Fox means to claim that truth and goodness, say, are simply cultural constructions that Fox identifies as "fantasies," fictions propounded to serve social or individual interests. No doubt cultures vary greatly from one another as to how to construe the difference between lower values and higher ones; as individuals do also. Some of those distinctions strike us as arbitrary. But *arbitrary* is not the same as *unreal* or *irrational,* whether the subject is values in general or ethical principles or the rules of a game. The question *Why?* can be put to all value claims and to all rules: why this rather than that? But that question is not necessarily an intelligible one, e.g., why is the distance between bases in America's true game 90 feet, the distance between pitcher's mound and home plate 60 ft. 6 in. Six inches? Why?

As I read Fox he does not intend to invalidate value distinctions as such. The claim, rather, is that value systems of present-day cultures are out of whack and have been so for a long (historical) time. They are out of touch with a fundamental and incontestable reality. Without exception, though in different ways, regnant cultural systems stand crossways to the natural order.

Diagnosis: The Brain Has Been Put into an Impossible Situation

As noted at the outset lament over loss of community is a poignant note in contemporary life. Fox thinks the note is important but largely misleading. It induces us to look in the wrong places for the causes of the loss of community and thereafter for such remedies as may lie within possibility.

So the initial fact is incontestable: community as a richly human "sociable" existence is long gone. Why, how, did this come to pass?

> We have stepped irrevocably outside the limits of our environment of evolutionary adaptation. (p. 22)

Let us move in on some specifics of this weighty charge. Item: biologically we are fitted out for one and only one kind of social environment.

> ... the brain, with its mechanical adjuncts ... is still the primate brain re-tooled by predation. And it is not an organ of cool rationality: it is a surging field of electro-chemical activity replete with emotion and geared for a particular range of adaptive responses. (p. 5)

Cro-Magnon knew this. His society was tribal based on blood kinship and undisturbed by illusions of a universal culture. But then things went wrong. Something like a Fall from primordial righteousness (innocence) occurs. Massive social organization (empire) draped with fraudulent ideology (religion giving way to philosophy) appears on the scene. And this aberration grows from strength to strength. Result, in a nutshell: dehumanization, deracination, homelessness, the submergence of *Gemeinde* (community) in *Gesellschaft* (society). This is where we are and what we are.

Thus we are confronted by another incontestable fact: the old (original) Adam will do tomorrow and tomorrow what he has always done throughout history: destroy the fake "homes" foisted on him. He cannot return to the original blessedness but he will not long endure any substitute. Sufficiently desperate or past all fertile caring he may drive the human enterprise irreversibly into the nuclear apocalypse. This would of course be the absolute nihilistic gesture: better an everlasting nothing than a pseudo-human existence. Perhaps with his last breath and moment he will thank whatever gods may be for a weapon absolutely certain to execute the suicidal wish.

There are several problematical features in this stunning diagnosis.

(1) Is the scientific picture true as well as conclusive? More specifically, what is known about the *human* life of Cro-Magnon? In that age (why not call it Golden and be done with it?) there were artistic geniuses in various places: witness cave-art. Is their work religious? Some scholars say so. That might suggest that a Fall has already occurred, namely, that the conditions of existence have become so problematical, if not flatly unendurable, that invocation of unearthly powers is mandated. For that matter, why should we not assume that in this respect Cro-Magnon is as human as Chicagoan?

(2) Consider the ancient and persistent metaphor of the Fall. Liberal Protestant theologians view the biblical case as non-historical. Traditionalist thinkers hold out for its historicity. Some post-liberals use the metaphor as a symbol, others as a non-historical category of history. But for our purposes the truly landmark figure in the construal of the biblical (OT) Fall is Augustine. He is this because of his mighty wrestling with the key question, How was it possible for the singularly endowed creature, Man, to boot away his perfect situation with God; and do this not of a moment's wild whim or irrepressible impulse but self-consciously? And almost as though he

had already developed the germ of a rationalization for this totally irrational act before he commits it?

Augustine's focus is the will of Adam (Man). So the question must be rephrased: How could the primordially perfect will of Adam have been moved to seek a good other than the perfect bliss of obediential communion with God his creator? If Adam's will were moved by an external power - whether physical-natural or diabolical - it would follow and it would be true that he was not responsible for his sin. On the other hand, if that fatal movement of his will were the effect of a flaw in his nature it would follow, again, that he would not be responsible for his sin since he did not choose or will his nature. It must be the case, then, that the first wayward movement of Adam's will, from which all human history derives (indeed *descends*) has *no* antecedent cause in the time-line. Thus, in respect to his sin Adam's will is a self-caused cause, a kind of unmoved mover. Not that Augustine uses that last phrase. I suggest that he comes close to it.

So on Augustine's reading our First Parent is in a remarkable philosophical fix. He cannot blame anyone or anything for his world-historical calamitous decision except himself. Genes, glands, social environment, toilet training or otherwise Mama, nasty old Id, Satan, God: all are home free; old Adam is stuck with himself as the explanation of his Fall.

Perhaps we ought to note in passing that Augustine ignores ignorance as an excuse. No Socrates he. Adam knew what he was doing. He knew he had lined up incommensurables and chosen and willed thereafter the lesser good.

Now what about Cro-Magnon, ourselves in primeval innocence? Why did somebody start yearning for the fleshpots of Egypt, so to speak, that is, for the comforts and power of empire? Did he sense at all that his imagination had produced a sick thing which if pursued would bring endless woe upon successive generations to the last syllable of recorded time?

These are not scientific questions. At best they are dubiously philosophical. So we must try to do better. Suppose then that the fatal transition was a forced option. Swelling population coupled with catastrophically reduced food supply required either a move to unknown, untried places or regimentation, perhaps even the creation of primitive class-distinctions. Apparently there is evidence that the ever-southward migration of prehistoric tribes in North America was caused by exhaustion of food supply. But why did Mayan and Inca and Aztec civilizations achieve such remarkable heights of social-organizational complexity and finesse, and superb aesthetic creativity as well, while others, both hunters and gatherers, lagged far behind in almost all categories? Who "fell"?

(3) The fundamental issue here is not whether Fox's diagnosis leads into some kind of Primitivism, though I think it does. The key question is how it is possible at this stage of history to assess the meaning of history's course with a cognitional organ, Brain/Mind, wrenched out of its biologically ordained environment. Add the further complication that the dirty work has been done by history itself, that is, by historical man to himself and thereafter to us all. Has the fallen creature any *real* memory of original (primitive) blessedness? In this respect biblical Adam seems quite well off. He remembers; this gives him an authoritative norm for assessing the quality of his life here-and-now. He knows where he is by where he isn't; who he is by who he once was. (Of course, this is what theologians have figured out for him.) Our natural inclination is to believe that such memories must be a curse. Adam should have been baptized in the river Lethe after he was kicked out of Eden. O if only God were fair!

The problems in Fox's diagnosis persist in his prescriptions for the pathological condition he has described and accounted for.

A Sound Prescription Bound to Fail

I want first to call attention to the tonality of this element in Fox's essay. We have already cited his use of "irrevocably." This is a reminder that there is no return to the true (=original) homeland of the psyche (Brain/Mind). Moreover, Fox does not expect contemporary cultures to heed the dire warnings, and venture to construct communities on a sound palaeolithic basis.

> I say it may be the best we can hope for; but I am totally pessimistic about its achievement. (p. 26)

But what is "the best we can hope for"?

> Perhaps what we really need in order to recreate the "communal good society" are organic extended kinship groups. (p. 10)

How to get from where we are to that condition? It will take some doing. The doing is "piecemeal social engineering" (p. 25). The goal is the restoration of the tribal ethos. The means are experiments scientifically inspired and controlled.

> We can attempt a realistic piecemeal social engineering of our over-individualized, over-rationalized, over-bureaucratized and altogether too large societies, to produce small-scale institutional settings where the palaeo-morality would have more chance to flourish. (p. 28)

Fox claims this is not a utopian solution. Moreover it could go wrong. The "organic-total" community could turn bureaucratic-totalitarian. It has done so over and over again. That is history for you. But why does this turn towards societal madness occur? Maybe Cro-Magnon really was a lame brain, didn't know what was good for him - and us. But just maybe he was already burdened with that self-involution of psyche out of which springs - the lie. That is Adam for you.

What then is this palaeo-morality from which history is so horrific a declension? Here Fox comes up thin. He gives us a delightfully poetic evocation of a life in that truly human era. But what are the *moral* elements in it? What of justice and beneficence? What of trust and truth-telling, of compassion and mercy? What of the modalities of love-agape and friendship as well as eros?

Fox acknowledges that time-honored religious teachers have had some things worthy of note to say about the moral order. Christ and Buddha and Mohammed and a company of lesser lights have offered some instruction on the right and the good. But *what*?

> ... regardless of their particular delusional systems (they) *have issued clarion calls for a return to palaeo-morality.* They have urged a return to the communal ethic of the tribelet. (p. 27; italics in the original)

He adds:

> Of the great three [Christ and Mohammed and Buddha], only Buddha seems to have been truly universalistic: the whole of mankind was his tribe. (p. 27)

This is an astonishing claim. Whatever Christians make of him Jesus is in the prophetic tradition of the religion of Israel. The highest reach of that tradition before Jesus can be seen in *Isaiah*: granted the special historic role of Israel, God's saving grace is extended to all the nations. As for the prior claim, the summons to "return to the communal ethic of the tribelet," I do not know what to make of it before Fox gives us some clues concerning the ethical coefficients of that mode of being.

So perhaps such claims would be more plausible if we knew the content of that palaeo-morality. And if we know also how any of it has managed to survive the Fall. But forget that miracle; I want to question his reading of the drive towards universality in the moral world. But not primarily the awarding of the dubious crown to Buddha. The prime issue in an universalistic ethics is not how much of humankind to include in/as tribe. *It is rather how other persons ought to be treated irrespective of their tribe.* In Kantian terms, persons as such are ends in themselves and ought never to be treated simply as means.

Was Cro-Magnon capable of understanding this and accepting it as "law"? Are we?

Apart from some such stipulations concerning the moral of such the celebration of community leaves a good deal to be desired. For otherwise it tends strongly, perhaps irrevocably, to identify a neighborhood and kinship ethos with community as a moral aspiration. Group solidarity is a precious commodity, no doubt. Jesus' disciples had it; and Al Capone's gang; and Robin Hood's; and Sigma Chi. We might even surmise that the marvelous artists of Lascaux were sustained by a great richness of fellow feeling. But that isn't what their art expresses. It speaks powerfully and authentically to us not because we are members of their tribelet but because we are human. Or because we aspire to that estate, and cherish all the help we can find to get us there.

I have no doubt that community rightly envisioned is a *sine qua non* of a life not only righteous but blessed as well. But so soon as we hear talk about community as kinship bonds some of us are reminded of a teaching: Who then *is* my brother? Who is my neighbor?

Theological Afterthoughts

Unity--Alienation--Reintegration. What an incredible "adventure" this triadic metaphor has had in reflection on the human condition, and, for that matter, on the cosmic process. Moreover, the career of this metaphor is not finished. Freud gave it new life, and deepened its mystery.

Much of the appeal of Fox's essay derives from his employment of this metaphor. Just as historically it emerged garbed in mythopoetic colors, and became a philosophical conceptual schematism, so Fox's essay is a warning that the reverse process is always a possibility: a primordial unity (community) may take on a mythic quality, and finally command a kind of reverence as the Omega as well as the Alpha of the human story.

There is a rather different problem in the employment of this metaphor. In American culture it collides with another one. And I think that this adversary has a deeper and stronger grasp of the American psyche than the ancient triad. This metaphor is also a complex one with these components:

(1) A deficient Given - home, tradition, culture.
(2) The lure of Novelty - unlimited possibility.
(3) Self-creation - the new being in a new world.

The stages (1-3) defy dialectical precision, the connections among them are historical, non-implicational. The sequences are non-logical, the latter moment is not an entailment of the earlier.

Oddities abound in this picture. Of these, few are more interesting than the configurations and vectors of *nostalgia*. The archetypical backward-longing is not for a "place" but for a mood. "Backward, turn backward, O time in your flight / Make me a child again just for tonight." We do not pine for the literal recovery of the old gang; at the end of the first hour of the Class Reunion we realize that wasn't what we came for. We seek rather for something like that powerful sense of a common life no longer corrupted by ego aggressions and ego capitulations. For "no longer" outranks in ethical seriousness any "not yet . . ." In religious terms, purity of will is the 'joy of our desiring,' not a magical retrieval of innocence.

Oddities notwithstanding, the dominant American metaphor does not incline us to take with ultimate seriousness any philosophy or social policy that urges a return to primitive community as our salvation. For the most part we cheerfully concede that we can't go home again. But who wants to? Who but the defeated, the lost and strayed and stolen? But even the most wretched victims of the social order do not ask for a return to an earlier condition. Let those who have had a home cry its loss. The underclasses long for something new. At our best so say we all.

But the longing for a novel community must cope with several beliefs (attitudes) so deep we might reasonably call them mythic.

One: *Bigness in social organization is badness.*

Something can be said for this philosophical neurosis, if it is said dialectically, that is in critique of Giantism. Surely bigger is not necessarily better. As surely bigger is not necessarily worse. True, massive institutions cannot be run as though they were Cub Scout dens. True, a metropolis cannot be loved as a neighborhood can be, though a 'successful' city is likely to be a nexus of neighborhoods.

Two: *Societal impersonality is the same as anti-personality*

This attitude has an important analogy in ethics, namely, the confusion of *disinterestedness* with *indifference* or simple lack of interest (concern). But caring equitably is hardly the same as not-caring. Impartiality in treating conflicting claims is not to be confused with indifference either to such claims or to a rational resolution of the conflict.

Nevertheless, in this second deep belief there is a tough problem: a strong antipathy for Bureaucracy. These powerful negative feelings do a great business with the metaphor of the Machine. Bureaucracies are machines. But communities are organisms. In a machine the relation of part to part is the same as the relation of part to whole: purely external; the parts have no "insides" and the whole system, no matter how complex, has no "soul." But in an organism the "parts" are constituent members bound together by internal relations; they are unified by a common pervasive life (spirit) that is greater than any member but is expressed in the decisions and achievements of every member. Thus every real member is irreplaceable. The organism would not simply perish with the death of a given member, whether of high degree or low, but it would not be the same whole. Reduced to a quasi-philosophical formula the difference between Machine and Organism is this: organisms remember; machines repeat.

This formula has an anthropic bias. In the American psyche bureaucracies are soul-less. Their functionaries follow the book: a rule is a rule; the book is, like God, no respecter of persons (individuals). The individual *per se*, whether as subject or object, counts practically for nothing. Whereas in the organism - the community - even the enemy's individuality is respected so far she/he belongs to it. All the great ambivalences are embraced by community. Passions are its lifeblood.

Looking beyond attitudes hostile to massive social organization we have to ask how is it possible to preserve or retrieve in the world created by modern life community as human enterprise deeper and richer than relationships in which the raison d'être is reciprocal utility?

This question has a philosophical ring to it but it is not a philosophical question. Nor is it a question amenable to resolution in the political arena despite appearances. It is now virtually *de rigueur* for politicians to espouse unyielding support for the traditional Judeo-Christian family, the family farm, village mores, *ad infinitum*. There may be some votes among these souvenirs. But how can such sentiments be translated into policy? To do so is to move inevitably towards the Machine. Let a congressman agonize ever so much over the predicament of John S. Bean on his farm north of Implausible, Nebraska (we must never forget that his great-grandfather fought and bled and died at Gettysburg in 18 and 63!) But a rational and humane and *possible* agricultural policy must prescind from that personal anguish. Passion may be the *motive* for change in social structure and process. *Intention* must reckon with conflicting interests and allocations of resources and effective machinery for administration of the policy. One must hope that the adopted policy will get there in time to save John Bean, and even put Implausible back on its economic feet. But if these good things do not come to pass, it is not necessarily the fault either of the policy or of the machinery fashioned to effectuate it. The fault, dear John, lies not in our machines but in ourselves....

Coda

Win lose or draw we are stuck for the time being with metaphors of religious provenance. Prof. Fox has not escaped this fate. His essay is infected, so to speak, with two of these metaphors, and they are monarchical: Fall and Paradigmatic Community.

Protagonists of the Fall do not agree on what or where or how come. The big point of this consensus is we are not where or what we once were (we = all members of the human community). The lost world may not have been perfect (but compared with what?) but it was a lot better than anything since. How better, in what particulars? Here the protagonists of the Fall are all over the playing field. Some say life then was far more natural, rife with spontaneous passion, uncomplicated by abstract rationality. Others put in a bid for innocence, a condition that is hardly a virtue but perhaps is better than that, namely a state of consciousness beautifully free of self-alienation effected by guilt. Others are beguiled (none more beautifully than Milton) by the Edenic myth, and make untrammeled communion with God, together with all the benefits thereof, the very definition of original community. But whatever the description of that blessed existence, whether or not blessed of God, we may be sure that whatever is appraised to be the harshest and deadliest pathology of human life as we now have it was absent before the Fall. *Absent by definition*: the postulation of pious belief, or of scientific hypothesis, or of poetic evocation. We cannot rule out a variety of combinations of these possibilities, e.g., Milton.

Paradigmatic Community. Here again a plethora of candidates is a mixed blessing: Epicurus' immured circle of high-minded friends; the early communist church in Jerusalem; the closed Mennonite community; New Harmony; Oneida; the Pig Farm; kibbutzim. Take your pick.

The ordained functions of the Paradigmatic community have also a splendid variety. I isolate a few of them arbitrarily: To save the world; to provide an asylum from an insane and/or sinful world; to show the world the true and certain shape of the future (practicing the Eschaton); to keep the faith no matter how the world behaves or is likely to wind up.

The ordained function of the Paradigmatic Community varies with the way the creative and sustaining power in/of the community is identified. Is that power of being exclusively human? Or is it exclusively divine? Or is that power both divine and human? By definition it has to be one or the other, or both; by definition that power cannot be satanic.

What is Prof. Fox's answer? The neo-palaeolithic community, that paradigm, will come to be by the operation of purely natural forces. Cro-Magnon is a product of natural forces under the "program" of evolution. His re-constitution must occur the same way. But is this what Fox means to say? It is to be doubted. But I don't know whether it is fair to suspect here something like Thomas Henry Huxley's famous positing of the ethical law in opposition to the law of nature. In more modest terms perhaps we should say that we cannot count on mother nature to extricate us from the awful mess we have created for and as ourselves. Since Cro-Magnon, if Fox is right, the course has been *devolution*. The high calling of the Paradigmatic Community is to reverse this process and put the human enterprise on the road towards redintegration with itself and with the natural environment.

But perhaps it is too late for anything but a symbolic gesture in that direction. Fox seems not to have much doubt about that; so my "perhaps" is not justified. The world will not be reformed by the Paradigm or by anything else. So why worry? The answer: let the world go its damned way; there is a far far better thing that persons of discernment and courage and goodwill can undertake. That is to develop a genuinely human pattern of association, like the extended family. Thus individuals can be saved, so to speak, that is, offered a context in which untrammeled individuality can be achieved without impoverishing or weakening the social fabric. In such a community all are enriched through the creativity of each participant in its life.

My largest reservation about this redemptive community, this paradigm of authentic humanity, has little to do with practicability. My overriding concern is the persistent vagueness of the ethical features of that community. Consider, for instance, justice. On this topic Fox is as silent as Marx. Like Marx Fox is outraged at the inhumanity of the present order. Like Marx Fox is silent or vague on how the blessed neo-palaeolithic community will honor the claims of justice.

I do not find that Fox's evocation of the Paradigmatic Community is clearer or more persuasive on other ethical principles.

Finally the Paradigm is perfectly short-suited on the religious side: it is blank. Perhaps he finds that the *religious* dimension of the great ethical teachers is, in his terms, delusional. These delusions have a long and deep history. They may have enmeshed Cro-Magnon himself. If that is so, then Fox should be grateful that the otherwise devolutionary course of human development has latterly shown us that in respect to religion Cro-Magnon cannot be our model. Indeed he has something to learn from Godless modernity. Isn't that progress?

A REPLY: BEARING THE BAD NEWS

Robin Fox

Faced with such thoughtful and forceful criticism, the temptation to reply at length and with elaborate and incisive self-justification is almost too much. But considerations of both time and space, to say nothing of the reader's patience, impose a merciful limit. I shall make a few points in reply to the particular criticisms I think most germane, and take the opportunity to make a final statement of my own.

First some disclaimers. I won't bother with many of the specific points raised that are, in Prof. Hartt's words, ground clearers. The reader can ponder them and make up his own mind. Also, on several of these issues (the brain/mind problem for example) I am on record elsewhere and there is no need to elaborate here. Nor will I attempt to respond to calls for more information on subjects that would require an encyclopedic response; e.g., requests for a description of palaeolithic society or a spelling out of the exact nature of palaeomorality. Again, I am on record with some of these, and as for the rest, only a lifetime's education in anthropology would suffice as answer. It is obvious that Prof. Handler has less problem here than the other two because he knows to what I am referring, and because he had the extraordinary good grace to obtain and read *The Red Lamp of Incest: An Enquiry into the Origins of Mind and Society*, (Notre Dame, 1983), where some of these things are set out at length. I could not spell them out in an hour lecture, and can't do so here.

That on one side, let me take up some issues that should be dealt with. One theme that runs through all three papers and that has me somewhat puzzled is the accusation of Rousseauianism. I thought I had made myself clear on this, but obviously not. Thus most of Prof. Klosko's paper and good deal of Prof. Handler's are devoted to exposing this Rousseauian tendency. The latter "wonders if" I cannot be placed in that pastoral tradition, and the former "was strongly reminded of" Rousseau's *Second Discourse*. Prof. Hartt does not specifically invoke Rousseau but uses the powerful image of The Fall to the same effect. Now I did of course explicitly distinguish myself from any "noble savage rhetoric" (Handler), and I tried hard to make clear that I was not advocating a "Golden Age" theory of any kind. But obviously I am hoist on the petard of my own love-affair with upper-palaeolithic culture. Despite my protestations that I am not saying that life was better or happier or more just, peaceful or healthy then, I am trapped by my insistence on giving the stone age a better press. What I wanted to say was that things were "better" for us then in the same way that we can all agree that life is better for an eagle free and on the wing than tethered in a cage. Even if the eagle only has a short and difficult life fraught with danger and adversity. It seems to me a simple argument so put. I dot not have to argue that the eagle is "happier"; happiness, as Mae West might have said, has nothing to do with it. All I am saying is that animals are better off in the untouched grandeur of their natural surroundings; better off in an objective not an evaluative sense; better off in that this is the niche for which they evolved and they cannot operate successfully outside it. Much of my paper was devoted to a demonstration that they do not operate well outside; cannot operate well; and will never operate well. This is I think only superficially Rousseauian. The selectivity in the view of the past involved in such judgments invoked by Prof. Klosko is certainly evident in all the Social Contract theorists. I had thought to be free of that by invoking a known past (partly, as he sees, through its analogues in the present) rather than the fictional one of Rousseau, Hobbes or Locke (or for that matter Lucretius or the Book of Genesis.) And when I say "known" I do not mean that all details are known, but that the general forms of society and culture are known, and that is enough to start with since I am only making a very general point.

This leads me to another point that runs through the papers. I am, they say, savagely critical of western rationality, and yet I seem to want a "privileged position for one product of western rationality, namely, evolutionary biology. Professor Handler does not think I can logically privilege it in this way, while Professor Klosko goes further to the point where we are devoid of any way of deciding on "the facts of the case" because there are no facts independent of evaluations. The latter position would leave us with no point in arguing at all, but it is a logical extension of the relativism invoked by the former. This is one of those argument like Berkeley's refutation of our knowledge of the external world which, while we can't accept it, is hard to refute. We are caught like the Cretan liar: if we say that all knowledge is relative, then our statement to that effect is relative and we have no means of establishing its truth. I can only try to sidestep this conundrum by saying that any relativism I might have invoked was a relativism of values (and not even that really since I am looking for a universal basis for values) but not a relativism of *truth*. I might think that *Homo sapiens* was "better off" in the palaeolithic, but I don't think his knowledge of the world could compare with our own. He had really very little notion of what he was or why he was what he was. (And this didn't matter one jot because he was doing just fine anyway.) Neither did we until relatively recently. But while I may not think that, in a human sense, living today compares well with living then, I would never want to say that the truths that our science reveals have no more validity than magical beliefs. By the very criterion Professor Klosko uses -falsifiability - they are superior in the information they convey about the external world. It seems to me that in this argument about the relatively of truth, one either accepts like Gellner that scientific rationality is not open to relativistic challenge or one is skeptical of such a claim. Klosko has the right to be skeptical, but I have the right to be a believer. Thus, I think that evolutionary biology (and adjunct information) can help us understand our situation better because it can ask the objective question "what is our environment of evolutionary adaptation" and give an answer based on hard scientific, refutable evidence; evidence that is, if not totally free of evaluative contamination, at least heading in that direction. To those however who are convinced of the relativity of all facts, this is no argument. But what would be?

And this leads me to the criticism that, according to Professor Klosko, I try to pass off a frankly evaluative position as "scientific." But I never said that the theory was "scientific." I explicitly state that it is "based on" a hard science view. It is of course an interpretation, as all theories are: in Popper's words, a "conjecture," and properly so. I was nurtured intellectually at Popper's knee (at the London School of Economics) and I well remember his stressing again and again that one does not "test or refute" theories as such; one tests or refutes hypotheses derived from them. All theories ultimately have the invulnerability that Klosko attributes to mine. In the end they are discarded because hypothesis after hypothesis is disconfirmed thus causing us to lose faith in them. If they are truly invulnerable then they are metaphysical statements not scientific theories. My theory has plenty of derivable and refutable (in principle) hypotheses. I state one quite explicitly: I predict that the post-industrial society will not work and will be destroyed by the very same factors that Bell (and Polanyi) identified as the ones undermining industrial society. Let us be patient and see whether or not this is disconfirmed. (It is strange to me that none of the commentators really took up this issue which to me is the heart of the paper. Perhaps if we'd had a sociologist it might have been different, but it would have depended on what kind of sociologist. One cannot be certain of such things in these days of intellectual flux and faddism.)

A couple more items. The human mind was invoked by two commentators, and Professor Handler made much of the "imaginative freedom" of the mind. My point here is that certainly the mind is free to invent new worlds; that is precisely what it does all the time. The question is, can we live in all of them or only in certain of them? That, I think, is the basic issue of the paper. It is not enough to say that we can create any symbolic system we like, of course we can; the question is can we apply it without destroying ourselves? Our record is not brilliant, and I have obvious grave doubts about our chances. On a technical note: I do not think that categories are so much *affected by* emotions (Handler) as that they *are* emotional by their very nature. The work on memory leaves us no other conclusion. Thus I think that Geertz, Lévi-Strauss and Handler are working on a totally wrong paradigm. To paraphrase Locke "There is nothing in the mind that was not first in the limbic system." (See *Red Lamp* chap. 7, and my "The Passionate Mind" in *Zygon*, 21 [1]: 1986). Also, I do not, along with other modern philosophers, accept the "fact/value distinction" as commonly understood (Klosko). It seems to me to rest on a misreading of Hume. As has recently been argued, and as Darwin saw quite clearly, it does not apply to functional statements. Thus, if a man *is* a navigator, it is perfectly logical to say that he *ought* to guide the ship: that is what navigators do. (Darwin's example was a pointer.) This makes "ought" a descriptive not an evaluative statement. Thus, it makes sense for me to insist that we "ought" to be human. The trick is to define the latter as objectively as one can define navigation or pointing. I have tried.

Again, I thought to distance myself from utopians but ended by being caught up in the powerful drive towards "reintegration" that Prof. Hartt so astutely recognizes as the end product of his interesting dialectic. We none of us like to be the bearers of bad news (*vide infra*), and hence I am always faced if by no one else by my children who, quite reasonably given that they are to live a long time in this vale of tears I am describing, demand that I offer some prospect, some hope, however dim. My feeble answer was to try to suggest what might, in a "best scenario" sense, be attempted. This is my Popperian piecemeal social engineering, small is beautiful, suggestion, and is I think quite deliberately anti-utopian. I don't hold out much hope for sweeping and dramatic solutions to all human ills as is the utopian way. I am only saying, "If you want to be a utopian, here is a possible formula that is less arbitrary in what it proposes than one based simply on humanistic - and (pace Handler) culturally relative - premises." But I perhaps should have left it alone. My "paradigmatic community" is, as Prof. Hartt says quite rightly (and this is echoed by the others), vague and impractical. Yes.

There are many other particular points I could discuss: the point about the prelapsarian Adam so beautifully made by Prof. Hartt and echoed by Prof. Handler, for example. Why did Adam/Palaeolithic Man push beyond his evolutionary niche? I did suggest "external" causes - ice and population pressure - because it seems to me that it could not have been his self-will or his capacity for symbolic thought *per se* that pushed him beyond. These were there from the moment he was *Homo sapiens* by definition, and in my estimation even earlier (*Homo erectus* had a more than respectable brain size well within the modern human range). Why then did he not make the move earlier than ten thousand years ago if that capacity was the *cause* of his move? But to spell that out would be several volumes again. There are other points. I do think Buddha was the more "universalist" - or rather that Buddhism was the more universal religion. But that would take some spelling out too. And it is not germane to the argument. But Jesus did say that his disciples should go first to the lost sheep of the house of Israel. And indeed the earliest proselytizing was in synagogues, and the first great debate was on whether or not to let in the uncircumsized and whether or not He was just the Jewish Messiah or the Saviour of the World. Paul may have clinched it for universalism, but this was a victory over the tribal tendencies. However, I could be wrong. And certainly Prof. Hartt is right to point out to me the difference between the real universalistic ethic and an ethic of universal tribal all-inclusiveness. I am grateful for the correction. There is a real difference. However none of this - or the many other specific points I might take up - really affects the argument, so I will leave it there with sincere thanks to the commentators for their serious and lively efforts to put me right. I am sorry to be so recalcitrant.

And here then I will venture the last word I promised. It has to do with the title of this reply: bearing the bad news. I have often argued and will repeat here that the modern social and political sciences cannot be properly understood unless their ideological program is understood. They are at heart utopian, carrying out the Enlightenment program initiated by Bacon and Locke, which would, through education and science, make human progress towards democratic, peaceful and prosperous societies inevitable. Indeed, when we think about it, this is so obvious as to be taken for granted. Why otherwise have social and behavioral sciences at all? Of course that is what they are for. Another view might be that what they are for is to discover the truth about human nature and human society whatever it might be and however damaging to our progressive, decent self-image. But for the majority of social scientists, as for the majority of people, this issue is settled. It is axiomatic that we can be improved and our societies improved; the only real questions are technical.

The result is that there is a tremendous bias in all the sciences towards the bearing of good news. It is inconceivable that any news refuting any part of the utopian program should be well received. For a start, the grants and funds would immediately dry up. No one in charge of spreading the bucks around the educational and scientific establishment is interested in the possible bad news that we are not perfectible, that our societies cannot be, in the end, improved beyond measure. Improved, that is, if only the social scientists could figure out how, which is their only real function. The bad news is usually delivered by renegade philosophers (Nietzsche,

Sartre), or by "humanists" (Orwell, Golding) or theologians of an orthodox stripe, who can be discounted because their opinions are not "scientific." H.G. Wells spent his long and active life dutifully delivering the good news about the possibilities of a scientific Utopia, but just before his death he wrote the remarkable *Mind at the End of Its Tether*, in which he concluded that "*Homo sapiens*, as he has been pleased to call himself, is in his present form played out." Imagine that as the basis of a research proposal. Or Orwell's proposition that the vision of the future was a boot stamping on a human face; or Sartre's that evil cannot be redeemed (*What is Literature?*); or Lessing's opinion quoted in my paper that we have very little idea what is going on, and what idea we have is largely erroneous. This worst of all.

Yet this other message has been with us since the Greeks and the Prophets and we should pay it some respect. Very few of us do or dare to. Like the Dean's wife with Darwinism, we hope that if it be true it not become generally known. Very recently a group of UNESCO-sponsored scientists issued a condemnation of biological views of human aggression precisely because they inspired "pessimism" and this would deter people from "peace activism." I find this a very silly argument, but a lot of eminent scientists - social and natural - signed it. (See my "The Seville Declaration" in *Academic Questions*, 1 [4]: 1988.) Given the nature of the Enlightenment project one can see why. Occasionally I find an anthropologist with the same information as mine, making a very similar point to the one in this paper, and am encouraged. Thus Melvin Konner in *The Tangled Wing* quotes Shakespeare, Henry James, Goethe and Malraux - all bearers of bad news, and continues:

> Let us invite these, as it were, artists of the soul to a cocktail party. On one side of the room are a group of tinkerers arguing cheerfully about various strategies for making everything just fine. On the other side, a group of biologists are discussing, rather glumly, the unchanging facts of human nature. Which group would they join?

There is little doubt where they would feel most at home. And mine is a version of the bad news: despite our innate flaws, we can flourish in a certain environment; but we have long ago abandoned the niche and in the new ones we create, our flaws get full destructive reign.

Despite having to face up to unpalatable truth, one always hopes. This is also very human, as the Greeks recognized in the myth of Pandora. And if the only hope we offer is that by facing the unpalatable truths we are better off than if we ignore them, then we will urge this morsel of wisdom on an ungrateful world. For even if we cannot achieve perfection, we can perhaps learn to live more easily with our imperfections. But this will never get the masses to the barricades, and certainly not get the tinkerers to cross the room to our side of the cosmic cocktail party.

MEMBERS OF THE COMMITTEE ON THE COMPARATIVE STUDY OF THE INDIVIDUAL AND SOCIETY 1987-88

Ravindra S. Khare CHAIRMAN OF THE COMMITTEE
Professor of Anthropology

Robert P. Scharlemann ACTING CHAIRMAN FOR 1988
Professor of Religious Studies

Theodore Caplow
Professor of Sociology

Ralph Cohen
Professor of English

Dante Germino
Professor of Political Theory

A.E. Dick Howard
Professor of Law

Robert Kretsinger
Professor of Biology

Robert Langbaum
Professor of English

David Little
Professor of Religious Studies

Eric Midelfort
Professor of History

Richard Rorty
Professor of Humanities

David Shannon
Professor of History

A. John Simmons
Professor of Philosophy

Kenneth Thompson
Professor of Foreign Affairs

Oscar A. Thorup, Jr.
Professor of Medicine

W. Laurens Walker
Professor of Law

THE SUBCOMMITTEE ON LEGAL CULTURES

Donald Black
Professor of Sociology

Ravindra S. Khare
Professor of Anthropology

Saul Levmore
Professor of Law

David M. O'Brien
Professor of Foreign Affairs

W. Laurens Walker
Professor of Law

COPY EDITOR AND COMMITTEE ASSISTANT - R.C. Alvarado

WORKING PAPERS SERIES

Other Working Papers published by the Committee on Comparative Study of Individual and Society include:

Working Paper No. 1: Critique of Modernity

>   Edited by Robert W. Langbaum
>   October 1986

Working Paper No. 2: Issues in Compensatory Justice: The Bhopal Accident

>   Edited by R. S. Khare
>   June 1987

Working Paper Number 3: Perspectives on Islamic Law, Justice and Society

>   Edited by R. S. Khare
>   September 1987

The camera-ready copy of this document was produced on a Compaq Deskpro 386/20 using WordPerfect Version 5.0. The font style is Times Roman and was printed on an Apple LaserWriter II/NTX.

"Desktop publishing" and layout assistance was provided by PHIL SCARR.

LIBRARY OF DAVIDSON COLLEGE